From Sensing to Sentience

From Sensing to Sentience

How Feeling Emerges from the Brain

Todd E. Feinberg

The MIT Press
Cambridge, Massachusetts
London, England

The MIT Press would like to thank the anonymous peer reviewers who provided comments on drafts of this book. The generous work of academic experts is essential for establishing the authority and quality of our publications. We acknowledge with gratitude the contributions of these otherwise uncredited readers.

This book was set in Stone Serif and Stone Sans by Westchester Publishing Services. Printed and bound in the United States of America.

Library of Congress Cataloging-in-Publication Data

Names: Feinberg, Todd E., 1952– author.
Title: From sensing to sentience : how feeling emerges from the brain / Todd E. Feinberg.
Description: Cambridge, Massachusetts : The MIT Press, [2024] | Includes bibliographical references and index.
Identifiers: LCCN 2024001005 (print) | LCCN 2024001006 (ebook) | ISBN 9780262550956 (paperback) | ISBN 9780262381468 (epub) | ISBN 9780262381475 (pdf)
Subjects: LCSH: Consciousness. | Neurobiology. | Senses and sensation. | Emotions.
Classification: LCC QP411 .F453 2024 (print) | LCC QP411 (ebook) | DDC 612.8/2—dc23/eng/20240314
LC record available at https://lccn.loc.gov/2024001005
LC ebook record available at https://lccn.loc.gov/2024001006

10 9 8 7 6 5 4 3 2

I dedicate this book to my wonderful wife, Marlene, who has been so supportive throughout our lives together. Also my gratitude to my daughter Rachel and her husband, Mike, and their kids, Jake and Kyla, and my son Josh and his wife, Rose, and their kids, Dorothy and Joni. Much love to you all for making this family so happy.

Contents

Preface

This book is about the origins and nature of *sentience*, the "feeling" aspect of consciousness, and especially about the role that biological and neuro-biological *emergence* plays in its creation.

As I explore in greater depth throughout this book, the idea of emergence is that *novel* features of a complex system may *emerge* as a result of the properties of the parts of a system and their interactions; and these emergent properties are not present in the parts of the system when these are considered in isolation. Thus, a "higher-level" property of a complex system has features that are more than the *simple sum* of the features of the "lower-level" parts.

As I discuss in later chapters, many theories of sentience and consciousness in general—especially those that are directed at solving philosophical problems—take a "top-down" approach wherein they consider sentience in a fully realized state in humans and then try to surmise how it is created. In this book, I take a more "bottom-up" approach in which I look for the origins of sensing in all living things and then attempt to figure out what is required for sentience to emerge from the parts of the system.

We will trace a journey of billions of years of biological evolution from the basic sensing capabilities of single-celled organisms such as the bacterium *Escherichia coli* all the way up to the sentience of animals with advanced nervous systems such as frogs, crabs, bumble bees, octopuses, velvet worms, and, of course, ourselves. In so doing, I will argue that we find no "magic bullet" that creates sentience nor any singular moment in evolutionary time when sentience emerged. Rather, we will follow a long and winding road of diverse and progressive emergent levels that culminate in what are the many sentient animals that remain on Earth.

To encapsulate and address the many biological, neurobiological, evolutionary, and philosophical problems that sentience poses, I present a new theory that I call *Neurobiological Emergentism* (NBE). The reason I chose to focus in this book on the emergence of sentience is that, in my view, emergent processes play a critical role in the creation of sentience and can also help unify some persisting biological and neurobiological issues regarding sentience with some perplexing philosophical questions as well.

First, NBE elucidates how neurobiological emergent processes operate generally in the creation of sentience; second, it elucidates how biological and neurobiological emergent processes run parallel with the evolutionary progression from *sensing* to *sentience*; and third, I believe that by viewing sentience as an emergent process, we can explain both its *personal nature* and the apparent *explanatory gap* between the brain and experience,

Plan of the Book

In chapter 1, I first outline in broad strokes what *sentience* is and how it relates to the more general term of "consciousness." I then consider some of the issues that make the scientific explanation of sentience and *feeling* so difficult. Here, I introduce the question of how and why sentience has a *personal nature* and discuss the philosophical question of what is referred to as the "explanatory gap" between the brain and sentience.

In chapter 2, I summarize the general features of biological emergence that will serve as a framework for my analysis of the progression from sensing to sentience. In chapter 3, I discuss the subtypes of sentience, and the criteria and evidence that can be applied to determine if an animal possesses sentience.

In chapter 4, I offer some general principles that can be used to trace the progression from sensing to sentience and how these features help us to distinguish nonsentient (simple sensing) from sentient animals. Here, I propose a three-stage model with a timeline that marks the progression from sensing to sentience.

The basis of the model is that sentience is a feature of *sufficiently neurobiologically evolved and complex brains*, and while it evolved from living single-celled organisms, it took as much as 3 billion years for sentience to emerge. Then, in chapters 5–7, I review the features of organisms and animals that fall within these proposed stages that lead to the emergence of sentience in the third stage.

In chapter 8, I discuss how NBE helps explain some of the conundrums that are faced when we attempt to relate neurobiological principles and mechanisms to the emergence of sentience and at the same time help explain how an "objective" (material) nervous system creates sentience with its many personal and "subjective" features. These two issues—the complex neurobiological emergence of sentience as well as the personal and unique character of "feeling"—have led some to claim that no *standard version* of the physical sciences could explain the emergence of sentience or, more generally, the emergence of consciousness.

One of these more extreme views has been called "strong" or "radical" emergence, a position that proposes that sentience and consciousness must be the result of some novel "fundamental" *physical property* or process. In contrast to this view, NBE is a "weak" emergence theory, the "weak" part meaning that standard and natural biological emergence processes are all that is required to explain the emergence of sentience from a sufficiently neurobiologically complex and evolved nervous system, just like all complex physical systems display emergent system properties.

In order to address these issues, I begin by discussing the position held by some that consciousness emerges "at the top" or at the "highest hierarchical level" of the nervous system or brain of a sentient animal; and that this emergence could be the result of some novel property of physics that uniquely results in sentience and the character of "feeling," a claim that is a version of a "strong" emergence theory of sentience.

To further explore this question and elucidate how sentience emerges in nervous systems in a manner that is consistent with standard emergent principles, I do a closer analysis of the many and unique diverse types of neurobiological *hierarchical levels of organization* that are characteristic of all sentient animals. I examine the role that neurohierarchical emergence plays in the creation of sentience and how it occurs across a wide range of neurobiological dimensions at both micro and macro scales, across different neural architectures and via various other distinctive neurobiological processes. The main point is that all of these processes create unique but standard *emergent system properties* that enable the "weak" emergence of sentience.

I conclude this chapter with a model of how these various emergent neurobiological processes interact and collectively create the *aggregate system
property* of sentience.

In chapter 9, I consider the relationship between NBE and some other
theories of or arguments about consciousness that most specifically relate to
it. These include *panpsychism, biopsychism* and the "emergentist dilemma,"
and *Integrated Information Theory* (IIT). Here, I discuss why sentience is an
emergent feature of life but life and basal sensing capabilities do not meet
the criteria for sentience, and that in the case of IIT, although this theory
in many respects differs from NBE, it actually lends support to some of the
emergentist postulates of NBE.

In chapter 10, I focus on the personal nature of experience and the
explanatory gap. In order to eliminate the seemingly unbridgeable gap
between the material brain and personal sentience, I will argue that just
like life itself is a personal emergent aggregate system feature of an embodied living organism, sentience is also a personal emergent aggregate system
feature of any particular organism—including ourselves; and, therefore,
the emergence of sentience can *only* be subjective and personal to that
organism.

Our understanding of the relationship between the personal experience
of a subjective *being* organism and its *objective* brain is critical for our understanding of sentience because the relationship between these two helps
explain one of the most controversial and still unresolved question regarding sentience and the explanatory gap: its personal nature.

I make the argument that in fact there is no "explanatory gap"; rather,
there is an *experiential gap* whose two elements—objective neural processes
and subjective experience—are both scientifically explained by standard
emergent principles. And therefore, ultimately, attempting to fully "objectify" the subjective *experiential aspects* of sentience is futile.

Finally, in chapter 11, I review and summarize the core postulates of
NBE and attempt to unite the principles of emergence in biology with the
particulars of the neurobiology of nervous systems, the evolution of sensing and sentience, and the aforementioned personal nature of sentience.

I summarize my view that the emergent nature of sentience is critical for
our understanding of what has been called the "explanatory gap." And one
important reason for this is that there are *two emergences* that create both
sentience and its personal nature and character. One is that sentience is an

emergent feature of neurobiologically evolved nervous systems as I describe in chapters 5–7; and second, that sentience is an *emergent feature of the individual life of the organism.* If we put these two emergences together—as occurs in evolution—we can rather simply see how sentience itself as well as its personal nature and character can naturally coevolve and coemerge from life and complex nervous systems.

Finally, I propose that rather than describing the problem of the personal nature of sentience as an *explanatory gap*, I suggest that this barrier to scientific explanation is better understood as an *experiential gap* between the objective knowledge of the brain and subjective first-person experience. I will examine the possibility that the "gap" between scientific or objective explanations of brain processes and subjective experience via "direct acquaintance" while real, *poses no scientific obstacle* to a full explanation of the personal nature of sentience and it can be naturally accounted for by the *emergent nature of sentience.*

Acknowledgments

There are many investigators and authors who helped with the development of this book. First, I thank Jon Mallatt who over the last decade has been my primary collaborator particularly with reference to the neurobiology and evolution of consciousness. I wish to thank the many scientists who took the time to answer my questions or review parts of the manuscript and were so helpful and generous in sharing their knowledge with me. These include Jonathan Birch, Lars Chittka, Andrew Crump, Robert Elwood, Robert Endres, Rhanor Gillette, Stevan Harnad, Claus C. Hilgetag, Jeremy Gunawardena, Thurston Lacalli, Christine Martin, Georg Mayer, Athanasios Metaxakis, Paul Nunez, Nicholas Strausfeld, and Nektarios Tavernarakis. And I once again thank my illustrator Jill K. Gregory, MFA, Certified Medical Illustrator, Associate Director of Instructional Technology, Icahn School of Medicine at Mount Sinai, and Past President, Association of Medical Illustrators, for her wonderful illustrations for this book. Finally, my deep appreciation to Phil Laughlin, Senior Acquisitions Editor at the MIT Press, for his support and guidance and helping in so many ways to make this book a reality.

1 Introduction: Neurobiological Emergentism

> But no matter how the form may vary, the fact that an organism has conscious experience at all means, basically, that there is something it is like to be that organism. But fundamentally an organism has conscious mental states if and only if there is something that it is like to be that organism- something it is like for the organism. We may call this the subjective character of experience.
>
> —Thomas Nagel, 1974.[1]

Something It Is Like to Be: Sentience

There are many words in the scientific and philosophical literature that have been used to describe the states of experience that are commonly referred to under the umbrella term "consciousness." But consciousness is a broad term that could apply to many diverse functions ranging from simple wakefulness all the way up to "higher states" of cognition and self-awareness. So first I want to make clear what I am trying to explain in this book.

Here, I focus on the *subjective aspects* of consciousness, its *feeling aspects*. While other terms such as "sensory consciousness," "primary conscious-ness," or "phenomenal consciousness" can be used to describe these basic experiential "feeling" states,[2] another term that may be applied more spe-cifically to the "feeling" aspect of consciousness is "sentience."

To be sentient specifically means "capable of feeling." The word is derived from the Latin *sentire*, which means "to feel."[3] It is this aspect of conscious-ness that Thomas Nagel, in the opening quote, suggested was "something it is like" for an organism "to be." And it is this aspect of consciousness that is one of the most difficult to explain.

Some writers have defined the subset of conscious experiences that are "sentient" as those that have a *valence* and as such are characterized by the positivity or negativity—especially the pain or pleasure—of these experiences. In this book, and more in line with Nagel's definition of *the subjective character of experience,* I will use the term "sentience" to cover *all* types of experience—whether sensory experiences from the world including what is seen, heard, tasted, smelled, and so on (*exteroceptive experiences*) or more internal experiences generated from the body (*interoceptive-affective;* pain and pleasure, emotions, etc.)—because all of these are *feeling* states that entail "something it is like to be" and, therefore, by that criterion, all experiences are sentient.[4] So while I may refer to "consciousness" when I am discussing other writers' works or opinions, in my theory of *Neurobiological Emergentism* (NBE), what I am trying to explain most specifically is sentience.

The Personal Nature of Sentience and the Explanatory Gap

Why Is Sentience So Difficult to Explain?
When we try to explain if and how there is "something it is like to be" a certain organism—Nagel chose a bat for instance—we are confronted with the problem of how certain brains are able to create "feelings." I think a key (perhaps the actual key) element that contributes to the "mysterious" nature sentience is its *personal nature.* This is because there does appear to be an "explanatory gap" between the *objective* biology of the brain and the inherent *subjectivity* of sentience. And in many respects, this is the most "mysterious" aspect of "consciousness" in general. This is one of the main questions that I focus on in this book, so first I provide a brief overview of that problem.

C. D. Broad and the "Mathematical Archangel"
Philosopher C. D. Broad (1887–1971) provided an early and, it turns out, prescient illustration of the apparent gap between the brain and sentience. Broad's full name was Charles Dunbar Broad. While he wrote widely in many philosophical domains, he is best known for his book *The Mind and Its Place in Nature,* which was published in 1925.

In *The Mind and Its Place in Nature,* Broad presented what would become one of the most famous thought experiments in the literature on the "mind-body problem." Broad's premise is simple: it involves the experience of the smell of ammonia. He argued that even if an omniscient "mathematical archangel" had total knowledge of the chemistry of ammonia and in addition also possessed full knowledge of the entire neurobiological basis of the smell pathway from the olfactory nerve to the brain, the archangel still could not predict the subjective *smell* of ammonia:

> He [the archangel] would know exactly what the microscopic structure of ammonia must be; but he would be totally unable to predict that a substance with this structure must smell as ammonia does when it gets into the human nose. The utmost that he could predict on this subject would be that certain changes would take place in the mucous membrane, the olfactory nerves and so on. But he could not possibly know that these changes would be accompanied by the appearance of a smell in general or of the peculiar smell of ammonia in particular, unless someone told him so or he had smelled it for himself. (Broad, 1925, p. 71)

The central point that Broad raises in this scenario is that no *knowledge* possessed by the scientifically omniscient archangel of the structure of ammonia and the brain could substitute for the actual *personal experience* of "what it is like" to smell ammonia. That no amount of explanation or analysis of the *objective* facts of the neurobiology of the brain can substitute for the *subjective* aspects of personal experience. It would appear that this poses a bit of a mystery for a scientific explanation of sentience.

Joseph Levine, David Chalmers, and the "Explanatory Gap"

Broad's view on the subjectivity of experience has been stated in various ways by many writers since he first made this argument. For instance, another influential way of expressing what is fundamentally the same problem was proposed by philosopher Joseph Levine in his 1983 paper *Materialism and qualia: The explanatory gap*. In that paper, Levine proposed that there appeared to be, what the title of the paper suggests, an "explanatory gap" between the *experiences* associated with sentience and the *neural processes* of the physical brain:

> However, there is more to our concept of pain than its causal role, there is its qualitative character, how it feels; and what is left unexplained by the discovery of C-fiber firing is why pain should feel the way it does! For there appears to be nothing about C-fiber firing which makes it naturally "fit" the phenomenal properties of pain, any more than it would fit some other set of phenomenal properties. The identification of the qualitative side of pain with C-fiber firing (or some property of C-fiber firing) leaves the connection between it and what we identify it with completely mysterious. One might say, it makes the way pain feels into merely brute fact. (Levine, 1983, p. 357)

And even more recently, in 1995, at a time when neuroscience was in full swing, David Chalmers also opined that Levine's "explanatory gap" was central to what he called the "hard problem" of consciousness. Chalmers contrasted what he referred to as the "easy problems" of consciousness—which is explaining the objective computational or brain processes that lead to consciousness—versus the "hard problem" of explaining subjective awareness. Here is how Chalmers characterized the explanatory gap:

> This further question is the key question in the problem of consciousness. Why doesn't all this information-processing go on "in the dark," free of any inner feel? Why is it that when electromagnetic waveforms impinge on a retina and are discriminated and categorized by a visual system, this discrimination and categorization is experienced as a sensation of vivid red? We know that conscious experience *does* arise when these functions are performed, but the very fact that it arises is the central mystery. There is an *explanatory gap* (a term due to Levine 1983) between the functions and experience, and we need an explanatory bridge to cross it. A mere account of the functions stays on one side of the gap, so the materials for the bridge must be found elsewhere. (Chalmers, 1995, p. 203)

The "Something Left Over" Argument

Therefore, the central problem that is raised by the aforementioned philosophers is that if you try to reduce or eliminate first person experience with an objective knowledge of brain functions, something is always "left out" of the explanation; namely, the "something it is like" part. Carruthers and Schier referred to this as the "something left over argument."[5] In their case, they relate the problem directly to the Chalmers "hard" problem of consciousness. Here is how Carruthers and Schier state it:

> The contrast between these two forms of the question lies at the heart of what we might call the 'something left over argument'. We have already seen this argument at work in characterizing the Hard Problem above. It can be summarized by the following slogan: 'no matter how many of the easy phenomena are explained there will always be something left over to explain, namely why experience is the way it is'. That which is left over is the Hard Phenomenon. The reason to claim this turns on the assumption that experiences are the wrong kind of thing to be approached with explanations that refer to the structure and function of mental states. (Carruthers and Schier, 2017, p. 70)

The Character of Experience

Another aspect of the personal nature of sentience that is even closer to Levine's explanatory gap is what Chalmers refers to as the problem of the *character of experience:* Why do brain states "feel" the particular way that they do? Why does "red" subjectively feel exactly and uniquely the way red does. Or, why does the activation of the auditory pathway lead to subjectively heard sounds? Isn't that beyond scientific explanation? Chalmers states it thusly:

> Why do individual experiences have their particular nature? When I open my eyes and look around my office, why do I have *this* sort of experience? At a more basic level, why is seeing red like *this*, rather than like *that*! It seems conceivable that when looking at red things, one might have had the sort of color experiences that one in fact has when looking at blue things. Why is the experience one way rather than the other? Why, for that matter, do we experience the reddish sensation that we do, rather than some entirely different sensation, like the sound of a trumpet? (Chalmers, 1996, p. 5)

So in related ways, these philosophers feel that there appears to be "something missing" when trying to explain how the brain creates the "feeling part" of experience. Certainly, at least historically, this particular issue poses one of the most significant obstacles for a scientific understanding of sentience. In order to "explain the gap," I propose that the processes of *biological emergence* hold the key to the solution.

Neurobiological Naturalism

So now I will explain how this book came to pass. Over the last decade, in order to address these questions and to better elucidate the nature and evolution of consciousness and, more specifically, sentience, I proposed, and then in collaboration with evolutionary biologist Jon Mallatt, elaborated upon a theory I called *Neurobiological Naturalism* (NN)[6] a name that was derived from John Searle's theory of *Biological Naturalism.*[7]

The main premise of Searle's *Biological Naturalism* was that all of consciousness and experience—including all "feelings"—can be explained as a "natural" biological feature of the brain that requires no new scientific laws or principles. As Searle put it:

> To have a name, I have baptized this view, "Biological Naturalism." "Biological" because it emphasizes that the right level to account for the very existence of consciousness is the biological level. Consciousness is a biological phenomenon common to humans, and higher animals. We do not know how far down the phylogenetic scale it goes but we know that the processes that produce it are neuronal processes in the brain. "Naturalism" because consciousness is part of the natural world along with other biological phenomena such as photosynthesis,

digestion or mitosis . . . Sometimes philosophers talk about naturalizing consciousness and intentionality, but by "naturalizing" they usually mean denying the first person or subjective ontology of consciousness. On my view, consciousness does not need naturalizing: It already is part of nature and it is part of nature as the subjective, qualitative biological part. (Searle, 2007, p. 329)

Searle also had a particular view of how to naturalize subjectivity. He called this perspective the "first person" or "subjective ontology of consciousness." His argument was that, as far as we now know, every state of consciousness is someone's or some animal's state of consciousness. As expressed in the following quote, his view was that this posed no obstacle to a complete *science* of consciousness:

Consciousness has a first-person or subjective ontology and so cannot be reduced to anything that has third-person or objective ontology. If you try to reduce or eliminate one in favor of the other you leave something out . . . biological brains have a remarkable biological capacity to produce experiences, and these experiences only exist when they are felt by some human or animal agent. You can't reduce these first-person subjective experiences to third-person phenomena for the same reason that you can't reduce third-person phenomena to subjective experiences. You can neither reduce the neuron firings to the feelings nor the feelings to the neuron firings, because in each case you would leave out the objectivity or subjectivity that is in question. (Searle, 1997, p. 212)

While we agreed in general with Searle's view that there was something inherently subjective about consciousness—what he referred to as its "first-person or subjective ontology"—it had one clear limitation regarding a clear biological explanation of the "something it is like to be" aspect of sentience, which really is one of the most—if not *the* most—mysterious aspects of consciousness. While Searle, in our view, rightly claimed that consciousness was *objectively* a natural biological feature of brains that can be explained by accepted biological principles, we still had no "biological" or scientific explanations for either the *emergence of consciousness* or its uniquely *personal nature* that I reviewed above. And it is these aspects of sentience that remain so refractory to explanation and one of the issues that I attempt to clarify in this book.

So, for instance, in contrast to Searle's examples of natural biological processes such as photosynthesis, digestion, or mitosis that have clear scientific explanations, sentience has a unique resistance to this sort of standard biological theorizing. In other words, we still didn't have a scientific explanation for the explanatory gap.

To address this issue, the theory of NN hypothesized that while the mechanisms that create consciousness indeed exploit some complex biological mechanisms that are *unique* to neurobiology, these are fully natural and do not entail any new scientific laws or principles. Second, as I elaborate further in this book, life itself was and is an important ingredient in the creation of consciousness as well as its subjectivity.

With these principles in mind, NN was based upon a roughly three-stage biological-neurobiological-evolutionary model that began with nonconscious single-celled organisms; an intermediate level of animals with more neurobiologically complex reflexes but not consciousness; and a final, third level that is reached by a diverse group of animals with neurobiologically more evolved brains and consciousness. This last stage occurred during the Cambrian period over 520 million years ago.[8]

Neurobiological Emergentism (NBE)

This brings me to the aims of this book. Over the last few years, it became increasingly apparent to me how centrally important emergence theory is for explaining *both* the neurobiology of sentience as well as its personal nature. Thus, here I present a theory that I call *Neurobiological Emergentism* (NBE), which is an outgrowth of and I believe an advance upon the theory of NN.

One of the main points of NBE is that it specifically emphasizes the role of *emergence* in explaining *sentience*. That said, it should be noted at the outset that the view that consciousness or sentience is an emergent brain process is not new. Indeed, the role of emergent brain processes in the creation of sentience has been frequently discussed and analyzed[9] and some of these are considered in greater detail in chapters 8–10. But I believe that my theory of NBE offers a new take on the relationship between and the integration of the biology, neurobiology, the evolution of nervous systems, and some unresolved issues regarding the philosophy of sentience.

The key principles of NBE are first, that just like life is an emergent feature of inanimate matter, that basic sensing capabilities that are present in all living things are emergent features of the life of the organism. Second, the progressive emergence of sentience from sensing can be traced to specific features of neurobiologically complex and evolved brains. Third, the version of emergence of sentience that I propose is not different "in kind" from the principles of emergence in general or in biology. Rather, I propose that it is the greater *degree of neurobiological emergence* in sufficiently evolved nervous systems that creates the emergence of sentience.

Finally, I propose that NBE not only helps clarify the biological and evolutionary progression from simple *sensing* to *sentience*; but at the same time it also explains how we can get from the *objectivity* of material brains to the *subjectivity* of experience that I mentioned above without any scientific or philosophical gaps.

This latter point is based upon on my view that the emergence of sentience *and* its personal nature cooccur and coevolved pari passu; that NBE can naturally explain both the biological basis of sentience as well as its personal nature without invoking any "mysterious," physically "novel" or "fundamental" properties; and rather than an "explanatory gap" between the brain and subjective experience that there is a nonmysterious and naturally occurring emergent "experiential gap" that poses no scientific barriers to our understanding of sentience. Thus, the theory provides a seamless explanatory pathway from the physical objective brain to the subjectivity of personal experience.

So, ultimately, in my view, attempting to fully "objectify" the *experiential aspects* of sentience is futile. However, I hope to show that NBE offers a fully naturalized theory of sentience that unifies the biology, neurobiology, and some aspects of the philosophy of sentience and consciousness without *denying* the personal subjective nature of sentience or trying to fully *objectify* it. Further, this emergentist position neither supports dualism[10] nor is it an argument against physicalism[11] and helps resolves some long-standing debates regarding the nature of consciousness and sentience.[12]

In the next chapter, I start by considering what I propose are the basic and general features that are essential to the emergence of sentience, the subtypes of sentience, and the criteria and evidence that distinguish sensing from sentience.

2 General Features of Biological Emergence

This appearance of new characteristics in wholes has been designated as emergence. Emergence has often been invoked in attempts to explain such difficult phenomena as life, mind, and consciousness.
—Ernst Mayr, 1982.[1]

In this chapter, I review the general features of emergence that in the following chapters apply to the biological and neurobiological emergence of sentience.

What Is Emergence?

The term "emergence" is derived from the Latin verb *emergo*. The emergo root means to arise or to come forth. Most sources on the subject trace its first use in the scientific literature to G. H. Lewes in his *Problems of Life and Mind* that was published in 1875.[2] There and ever since, it has basically meant that novel features of a system may emerge as a result of the parts of the system and their interactions; and these emergent properties are not present in the parts of the system when those are considered in isolation. It is commonly stated that an emergent "higher-level" property of a complex system has features that are more than the sum of the features of the "lower-level" parts.[3]

General Biological Emergent Features

Since my primary interest here is in explaining the emergence of sentience in living organisms, this book focuses on the emergent biological properties

Table 2.1

Major general features of biological emergence

• An emergent feature is an *aggregate system feature* that is not present in the parts. Thus, emergent properties are *novel* in comparison to the properties of the parts that create them.
• In order for emergence to occur as an aggregate system feature the individual parts of the system must be *physically united, integrated, or at a minimum interacting in some fashion.*
• Emergent features are *processes* created by the dynamic interaction of the system's parts.
• *Hierarchical systems* are critical to and increase emergent properties.

(Table adapted from Feinberg and Mallatt, 2020)

that are commonly found in all living things, and especially features such as hierarchical arrangements that play a particular role in the creation of sentience (table 2.1).

That said, the principles of *biological emergence* enumerated below are consistent with the features of emergence in general. But since emergence in biology has certain specific features, when I refer to emergence generally, I am referring most specifically to emergence in biological systems. Thus, they apply to any living plant or animal down to single-celled organisms.

Aggregate System Functions and Novelty

First and foremost, emergent features are "higher-level" properties of a complex system that are created by the collective functions of that system's parts that the individual parts in isolation do not possess. As noted earlier, it is often said that an emergent feature is "more than the sum" of the system's parts and hence an emergent property is *novel system feature* that is not "reducible" to the individual parts that create it.[4]

A popular example of an inorganic system with emergent properties is water, which has the emergent property of liquidity whereas no single water molecule has the property of being "wet." But more importantly, an emergent feature that is more relevant with reference to the creation of sentience is "life." Here, an atom or a DNA molecule within a living organism is not "alive," but single-celled organism or a collective body that is comprised of single cells are alive.

As a result, there is general agreement that emergent properties by definition are *novel* when compared to the properties of individual parts that create them. Philosopher Elly Vintiadis provides a concise summary statement that addresses the novelty of emergent features in relationship to the aforementioned characteristics of aggregate system features:

> If we were pressed to give a definition of emergence, we could say that a property is emergent if it is a novel property of a system or an entity that arises when that system or entity has reached a certain level of complexity and that, even though it exists only insofar as the system or entity exists, it is distinct from the properties of the parts of the system from which it emerges. (Vintiadis, 2013a, p. 1)

The proposal that an emergent property entails an aggregate system feature that is "distinct from the properties of the parts of the system from which it emerges" is clear. But the more interesting question when it comes to the emergence of sentience is *what type* of emergent novelty does sentience represent, and this question gets back to the problem of "weak" versus "strong" emergence that I mentioned in the Preface. This issue is specifically addressed in the later chapters of this book, but as stated earlier, I advocate a "weak" emergentist view.

Unity

In order for biological emergence to occur, the parts that contribute to an aggregate system must be physically connected, integrated, or at a minimum interacting in some manner. The degree of the connection between the "parts" can vary greatly depending in part upon whether the emergent feature is within or across hierarchical levels. Even in nonneurobiological hierarchies, for instance in the liver, the aggregate emergent functions performed by "lower-level" individual hepatocyte cells rely upon much closer physical and temporal continuity between the cells when compared to the emergent functions of the larger lobule of which they are a part.

Unity is a particularly important feature of emergence when attempting to explain sentience. Most obviously, while the nervous system as objectively analyzed is comprised of many billions of individual neurons, sentience is experienced as a largely unified field.[5] This sort of emergent integration would not be possible if the parts weren't in some way connected.

Process

Emergent features are *processes* created by the dynamic interaction of the system's parts. While the neuroanatomical, objective, material "parts and wholes" of a nervous system are in some ways the easiest to define, sentience is as much a process as Searle's examples of other biological processes mentioned in chapter 1 such as digestion, photosynthesis, or mitosis. The *dynamic interaction* of the nervous system's parts that are processes are also critical to explaining the emergence of sentience. This is also one reason why it is not possible to strictly *localize* an emergent feature such as sentience in the same way that we can localize any given anatomical feature of a nervous system.

Hierarchical Systems Promote Emergent Properties

Hierarchical arrangements, and especially neurohierarchical connections and processes, are absolutely essential for the creation and emergence of sentience.[6]

The key features of emergence in biological hierarchical systems are listed in table 2.2.

As noted above, as is the case with emergence in general, in order for emergence to occur as an aggregate system process, the individual parts of the system must be interacting directly or indirectly in some way. In neurobiological hierarchies "lower levels" often influence "higher levels," which then in turn influence the lower levels, and structures within the same level also influence each other via extensive reciprocal connectivity. We will see how reciprocal connections are exponentially increased in advanced brains with emergent sentience.

As I address at length in chapter 8, another virtue of neurohierarchical arrangements is that emergence may occur *simultaneously* at multiple spatial

Table 2.2
Major features of emergence in biological hierarchical systems

- *Hierarchical arrangements* are important for the creation of emergent features in all of biology. This is especially true of neurohierarchical systems.
- *Reciprocal connections* among parts within and between levels of biological hierarchies greatly enhance the emergence of novel properties in biological and neurobiological hierarchies.
- Biological emergent properties may occur *simultaneously* at multiple spatial and scalar levels and across diverse temporal frequencies.
- *Novel properties* emerge in a *system as a whole* as additional (typically "higher') levels are added.

and temporal scales. They may also have *diverse neural architectures* so that some may be *nested* in which higher levels are comprised of lower levels (e.g., lower levels are nested within upper levels) and others are *nonnested,* in which lower levels project their neural activation and "information" to higher levels, so that the lower levels are not "physically contained" within higher levels. All these pathways enhance the overall emergence within the entire system.[7]

Also, as the hierarchical system evolves and if more levels are added, as typically occurs in biological and neurobiological systems, there is more *specialization of its parts and levels*, both structurally and functionally. For instance, in fish, the visual system is centered in the tectum, but via the evolution of the visual system moves rostrally toward the cortex in mammals.[8]

Summary and Importance of General Emergent Features

The analysis of the general features of emergence reveals several notable points of interest regarding the emergence of sentience. First, all animals, whether they are—in my view—sentient or not, possess most of the general features of emergence that are present in all biological systems. But I will argue that sentient animals are distinguished by their many unique neurobiological features, paramount among them being the presence of specialized neurons and multiple levels and hierarchically interconnected neural hierarchies that advanced central nervous systems provide and make possible the creation of novel emergent properties.

3 The Neurobiological and Behavioral Criteria for Sentience: Subtypes and Supporting Evidence

In this chapter, I outline the features of brains that across various criteria can—in my view—be reasonably hypothesized to be present in animals that are sentient.

First, at the outset, it must be acknowledged that currently there are no absolute and universally agreed upon evidence or criteria for objectively determining the presence of sentience. This is especially true in the case of animals other than ourselves.

In the philosophical literature some aspects of this conundrum are commonly referred to as the "other minds problem" and it has a long and complicated history.[1] In any event, one thing is certain. Since we cannot (at least currently) ever actually *experience* another person's or another organisms' "mind"—for instance "what it is like to be" Nagel's bat, much less if there is "something it is like to be" an *Escherichia coli* bacterium that I discuss in chapter 5—when making a judgment as to whether sentience is present, we need to rely on inferences that are based upon a set of general criteria for sentience and the supporting evidence in any given case.

And to make matters even more complicated, sentience itself occurs in different subtypes, and these varied manifestations carry with them potentially different associated evidence or criteria. Thus, we need to utilize some general guidelines and hypotheses based upon a diverse range of parameters to *infer* whether there is "something it is like to be" in any given case.

General Principles for the Emergence of Sentience

Subtypes of Sentience

For the purposes of determining the presence of sentience, I find it helpful to divide the problem into three main issues. First, what are the major subtypes of sentience; second, what is the requisite *neurobiological infra-structure* of these subtypes; and third, what is the *behavioral evidence* that could support the presence of these subtypes.

First, I discuss some partially overlapping subtypes of sentience in con-junction with their neural infrastructures. In the emergentist view pre-sented here and elaborated elsewhere,[2] sentience encompasses *all* types of subjective "feeling" experience and these can be broadly divided into two types: sensory feelings that are generated from the world (*exteroceptive sen-tience*) and those that are more internally generated (*interoceptive-affective sentience*; pain and pleasure, emotions, etc.).[3] This distinction is impor-tant because some of the neurobiological and behavioral evidence differ depending upon the subtype under consideration and for the hypothesized role of emergence in the creation of sentience in general.

Exteroceptive Sentience: Mapped Neural Representations and the Emergence of Sensory Mental Images

Exteroceptive sentience is defined as the capacity of a particular species, class or emergent stage of organisms to display not only basic *sensing* capa-bilities (for example the capacity to sense a feature of the environment via vision [photosensation], mechanical sensation [touch and proprioception], chemosensation [including taste and smell], etc.) but the organism also has "feelings" associated with these sensing processes. And one important fea-ture that helps to distinguish simple *sensing* of the environment from *sen-tience* is whether the sensory processes in question are able to create *sensory mental images*.

Indeed, Gerald Edelman made mental images a critical feature of what he called "primary consciousness," concluding with: "Primary conscious-ness is the state of being mentally aware of things in the world—of hav-ing mental images in the present."[4] In a similar fashion Antonio Damasio[5] discusses how "mapped neural patterns" can result in "mental images" that are a part of what he calls *core consciousness*, which is fairly similar to Edelman's "primary consciousness." These exteroceptive "feelings" such as visual images, sounds, touch, smells, and tastes represent a critical part of sentience and "something it is like to be." So the creation of sensory mental images is a good marker for this form of sentience.

So the next question is what kind of evidence would be required for or support the presence of sentient sensory mental images? The primary evidence for the presence of an exteroceptive mental image is that we know that in order for the brain to create a unified sensory mental image, there must be some form of higher-order *mapped neural representation* of that sensory domain in the brain. This is a necessary neurobiological feature that helps us make the distinction between basic sensing and sentience.

Mapped neural representations actually come in various forms depending upon what sensory domain is involved. For example, one well-known type of mapped neural representation is a *topographical* map. A topographic map means that *spatial* ordering is preserved from the lowest level of the sensory field to the higher levels in the central nervous system (*topo* = spatial map). That is, the same, precise organization of neurons and their synaptic connections characterizes all levels of the neural hierarchy and match the features the spatial arrangement of the original sensory receptors.

A classic topographic map is the *somatotopic* map for the touch-related senses (e.g., light touch, position sense, nociception, etc.) that roughly preserves the spatial location of a stimulus on or in the body with its processing position in the nervous system and brain.

In humans, the skin and the rest of the body surface are spatially represented point by point, although in a distorted way. Especially large are the representations of the most touch-sensitive parts of the body—the face and hand—where the most sensory processing occurs. Another classic example of topography is the *retinotopic* map, which in mammals is located in the back of the cerebral cortex in the occipital lobe. It is here that the visual field is mapped and preserves the same spatial relationship that is found in an animal's retina.

However, not all these brain maps reflect the locations of stimuli in physical space. For instance, the auditory *tonotopic* map is spatially organized according to sound frequencies and the chemosensory maps senses of taste and smell are primarily organized around chemical receptor properties (Figure 3.1).[6]

These mapped representations are important for our understanding of the emergence of sentience for a couple of reasons. For one, the presence of a mapped sensory representation is not in and of itself sufficient evidence for exteroceptive sentience. For instance, there are somatotopic maps in the earlier processed and lower neurohierarchical levels in the spinal cord that are part of the sensory pathways for sentient touch but these mapped representations *do not in and of themselves create unified, sentient mental images.*

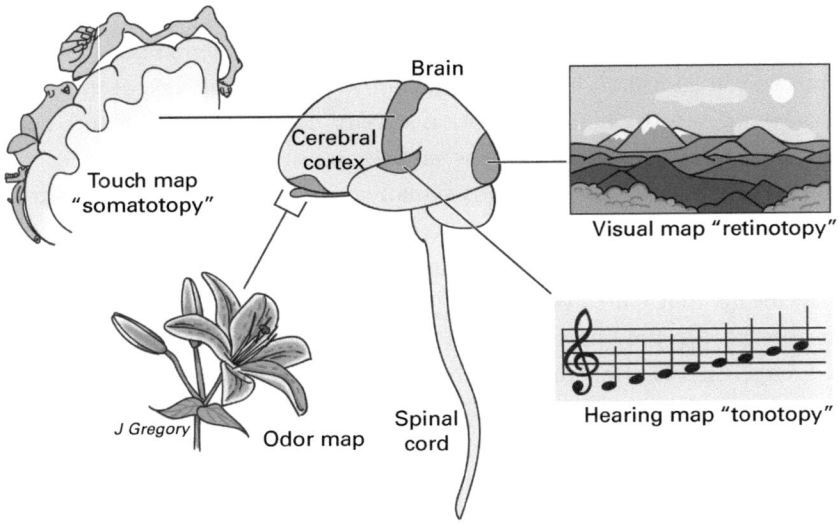

Figure 3.1
Exteroceptive topographical maps and "mental images." One indicator of sentience—in this case exteroceptive sentience—is that an organism displays not only basic sensing capabilities such as chemosensation (including taste and smell), mechanical sensation (touch and proprioception), hearing, vision, and so on,etc. but in addition it is able to create sensory mental images based upon these capacities. These metal images rely in part upon the brain's capacity to create relatively unified higher-order mapped neural representations.

The latter require successively hierarchical mapped representations that I discuss in the following chapters.

But from the standpoint of the emergence of exteroceptive sentience, these lower-level maps can serve as instructive neurobiological markers for the transition from the simple sensory processes that occur at lower neuro-hierarchical levels that are on the path to the exteroceptive sentient *sensory mental images* that are—I propose—present in all animals with this form of sentience.

Thus, from the standpoint of the NBE view presented here, we can trace the natural progressive *neurobiological emergence* of exteroceptive sentience along with the increase in neurohierarchical organization provided by more advanced brains (table 2.2). I discuss this progression and the emergence of sentience in more detail in chapters 4–7.

Interoceptive-Affective Sentience

A second type of sentience is the presence of *interoceptive* and *affective* feelings. While exteroceptive sentience is concerned with the processing of information from the external environment, these sentient feelings are more internally derived. Interoceptive feelings are therefore more responsive to physiological changes in the body and include sensations such as pain, hunger, thirst, motivational drives, and so on, while affective awareness is a broad category that covers all emotional (valenced) experiences.

Exteroceptive and interoceptive-affective sentience need to be considered in tandem as well as separately because there are some important differences between the neurobiological bases of theses subtypes of feeling. For instance, in contrast to the tight mapped neural representation that characterize exteroceptive awareness, interoceptive-affective experiences do not require this sort of succinct neural mapping and their underlying neurobiology involve some different anatomical structures (figure 3.2).

Another significant difference between exteroceptive versus interceptive-affective feelings is that while exteroceptive sentience does not *in itself* carry positive or negative emotional feelings to its sensory images, interoceptive-affective sentience does. Thus, sentient interoceptive-affective experiences are said to have *valence*.[6] Here is how Adolphs and Anderson define valence:

> Valence is thought by many psychological theories to be a necessary feature of emotional experience (or "affect"). It corresponds to the psychological dimension of pleasantness/unpleasant, or the stimulus-response dimension of appetitive vs. aversive. (Adolphs and Anderson, 2018, p. 66)

In another valuable review on the subject of affective valence, Kent Berridge breaks affective valence into two components: the *emotional aspect* or what he calls positive hedonic (pleasure) or negative hedonic (displeasure or pain) and the *motivational aspect* (approach or avoidance of a stimulus.[7]

> The hedonic aspects of affective valence include positive hedonic impact (pleasure) or negative hedonic impact (displeasure or pain). These are the affective kernels of rewards and punishments as well as components of many emotions. The motivational aspects of affective valence include functions that promote pursuit of rewards (such as incentive salience and declarative goals) and functions that mediate threat avoidance (such as fearful salience and passive avoidance). (Berridge, 2019, p. 225)

I believe that Berridge makes a useful distinction here that is relevant for the determination of the presence of interoceptive-affective sentience

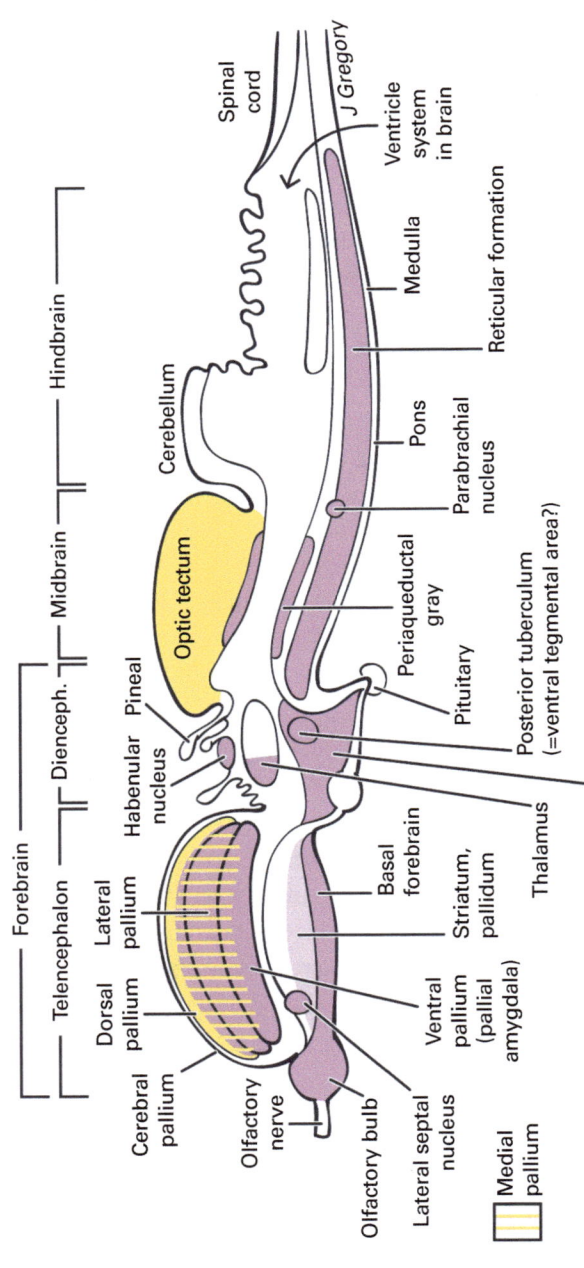

Figure 3.2

The structures involved in exteroceptive versus affective sentience in a basal vertebrate brain. Yellow regions are primarily involved in the creation of exteroceptive higher-order *mapped neural representations* and *sensory mental images*; purple regions are primarily involved in the creation of *affective sentience*. Cross hatched regions are involved in *both*. (Adapted from Feinberg and Mallatt, 2018a.)

that are considered next. This is because—at its most basic *sensing* level—the capacity for a stimulus to evoke the tendency to *approach* (positive-appetitive) or *withdraw* (negative-aversive) is universally present in all living organisms since these are necessary for feeding or to avoid bodily harm in any form, be it a noxious stimulus that could result in cellular damage or any other potential threat to the well-being of the organism.[8] But I will argue, as have others (see further discussion in later chapters), that these simple sensing behaviors, although they appear to be precursors of sentience, can occur without sentience and therefore are not evidence of sentience.

Evidence for the Presence of Interoceptive-Affective Sentience

So what constitutes evidence for both the emotional aspects of sentience such as felt pain beyond nociception (simple reflexive reaction to a noxious stimulus) or the presence of affective sentience in general? This may be judged in several ways.

One is the presence of a *neural infrastructure* that is capable of producing interoceptive or affective sentience that are summarized in figure 3.2. This neuroanatomy is also discussed at greater length in chapter 7. A second type of evidence is *behavioral* in which it is inferred that an organism is experiencing sentient (valanced) feelings.

An excellent compilation of both neural and behavioral evidence or what they refer to as "criteria" of interoception and affect comes from the recent work of Birch et al. and Crump et al. (table 3.1).[9] While their criteria were initially designed to assess *pain sentience* in species such as cephalopod mollusks—for example the octopus—and decapod crustaceans including crabs (animals that are discussed later on in chapter 7), they can also be applied to assess pain sentience across species.

I find these criteria for evidence of pain sentience succinct, convincing, and comprehensive, and they are also consistent with the prior neurobiological-evolutionary analyses that Mallatt and I did.[10] Thus, while Birch et al. and Crump et al. were primarily directed at providing evidence for sentient pain, these principles can be applied as evidence for sentient valance and motivation for interoception and affect in general. Finally, for present purposes, I find them particularly useful for both *distinguishing* simple sensing from sentience as well as tracing the *emergence* of sentience from sensing

Table 3.1

The Birch et al. criteria mainly for pain sentience, adapted from Birch et al. (2021) and Crump et al. (2022a; 2022b)

1. *Nociception.* The animal possesses receptors sensitive to noxious (i.e., harmful, damaging) stimuli (nociceptors).

2. *Sensory integration.* The animal possesses brain regions capable of integrating information from different sensory sources.

3. *Integrated nociception.* The animal possesses neural pathways connecting the nociceptors to the integrative brain regions.

4. *Analgesia.* The animal's behavioral response to a noxious stimulus is modulated by chemical compounds affecting the nervous system from either or both an endogenous neurotransmitter system that modulates pain or distress or by local anesthetics (e.g., anxiolytics or antidepressants).

5. *Motivational trade-offs.* The animal shows motivational trade-offs, in which the negative value of a noxious or threatening stimulus is weighed (traded-off) against the positive value of an opportunity for reward, leading to flexible decision-making.

6. *Flexible self-protection.* The animal shows flexible self-protective behavior (e.g., wound-tending, guarding, grooming, rubbing) of a type likely to involve representing the bodily location of a noxious stimulus.

7. *Associative learning.* The animal shows forms of associative learning in which noxious stimuli become associated with neutral stimuli, or in which novel ways of avoiding noxious stimuli are learned through reinforcement. These forms of associative learning go beyond classical conditioning in which a single conditioned stimulus overlaps temporally with an unconditioned stimulus. Note: Forms of associative learning that are linked, at least tentatively, to sentience in humans (such as instrumental learning, reversal learning, and trace conditioning) provide stronger evidence than other forms.

8. *Analgesia preference.* Animals can show that they value a putative analgesic or anesthetic when injured by at least one of the following: the animal is able to learn to self-administer putative analgesics or anesthetics when injured; learns to prefer, when injured, a location at which analgesics or anesthetics can be accessed or prioritizes obtaining these compounds over other needs (such as food) when injured.

and I will also apply many of their criteria for sentience going forward in chapters 5–7.

So, as I outlined above, one way of breaking down the various lines of evidence is by judging whether there is *neuroanatomical* and/or *behavioral* evidence for interoceptive-affective sentience.

Following these guidelines, we see that Birch et al. and Crump et al. criteria 1–3—*Nociception, Sensory integration,* and *Integrated nociception*— are neuroanatomical evidence, while their criteria 5–7—*Motivational trade-offs, Flexible self-protection,* and *Associative learning*—are primarily behavioral evidence. Criteria 4 and 8 *Analgesia* and *Analgesia preference* combine neu-rochemical and behavioral features; therefore, for behavioral features, I will apply criteria 4–8.

Why Are Features 4–8 Behavioral Criteria for Sentience? Reflexive versus Nonreflexive Pain-Related Behavior

I will have more to say in chapters 4–7 about what neural infrastructure is required for the emergence of sentience. But first, I want to touch upon a behavioral variable that ties together the seemingly diverse behavioral criteria.

Criteria 4–8 are designed to be able to *infer* an animal's *experience* of pain. And one important behavioral variable for judging the presence of interoceptive-affective sentience that is, from my perspective, especially relevant for the neurobiological and evolutionary transition from sens-ing (simple nociception) to sentience is the relatively *nonreflexive nature* of sentient-indicating behaviors when compared to simple sensing.

This is because all that is required for nociception (from the Latin *nocere*, to injure) is the simple sensing of a noxious stimulus (e.g., mechanism, ther-mal) that could cause injury to the organism and the reflexive withdrawal responding that noxious stimulus. In contrast, nonreflexivity is a factor in all their behavioral criteria for determining the presence of pain sentience. So, what do we mean by nonreflexivity?

Here is a nice definition of a reflex from Robert Elwood:[11]

> A reflex is a simple, short-term response so any complex long-term response can-not be regarded as reflexive, rather it demonstrates an extended alteration of motivational state. (Elwood, 2019, p. 2)

As we will see in chapter 5, even single-celled organisms are capable of nociception, which is the simple detection of potentially injurious stimuli that is usually accompanied by a nociceptive withdrawal reflex away from that stimulus. However, for pain, there must be nonreflexive conscious awareness of the experience. In other words, the nonreflexive nature of a behavior pertains to a judgment as to whether an organism or animal is simply responding reflexively in an adaptive way to a given stimulus or state (*sensing*) or whether there is the presence of *sentient* ("felt") volition, intention and—as captured by Berridge and Elwood in the above quotes—*motivation*. Thus, the presence of nonreflexivity is another necessary crite-rion that helps distinguish basic and adaptive reflexive *sensing* that occurs in all microorganisms and animals from *sentience* that is a higher-order nonreflexive emergent feature of neurobiologically complex brains (see chapter 4).

Although it is not always easy to draw a strict line between reflexive and nonreflexive behavior, in general reflexes are faster, more hard-wired, automatic, and stereotyped, and more likely to be innate and not require learning. Also, reflexes in general are less neurohierarchical than nonreflexive behaviors and tend to not require higher-order brain structures beyond, for instance, the spinal cord and brainstem. So, in general, the more neurologically complex and neurohierarchically based a behavior is, the less likely it is reflexive and more likely it is to be sentient. This is why typically, by definition, reflexes do not require consciousness.[12]

One common example of a relatively complex reflex is the pupillary light reflex that occurs when a bright light is shined into the eye. The light stimulus—an exteroceptive stimulus—enters the pupil and, in sequence, reaches the retina, the retinal ganglion cells, the optic nerve, and then the midbrain. Here, reflexive and neurobiologically hard-wired outflow connections eventually cause the pupil to constrict and automatically adjust the eye to the entering light stimulus. This is a fast (measured in milliseconds), involuntary, innate, and nonconscious neural response that will occur even if an animal or person is asleep or even in a coma.

The same is true of the knee jerk that is also a response to an exteroceptive stimulus. In humans, the knee jerk requires just two neurons—one sensory and one motor—that are connected by a single synapse. And as the case with the pupillary light reflex, no sentience is required for this reflex to operate.

I raise this point here, and as I will examine in greater depth in the following chapters, because many behaviors that are present even in single-celled organisms such as the bacterium *E. coli* or the ciliate *Stentor roeselii* that involve seeking food or avoiding noxious stimuli are surprisingly complicated but—in my view—are sensing but lack evidence of sentience. Indeed, the apparent complexity of the sensing adaptive behaviors of single-celled organisms that may *appear* to be sentient has led some scientists and writers to argue that these behaviors are nonreflexive, intentional, and indicative of sentience. However, I believe that these behaviors are reflexive and these organisms are not sentient.

Along this line of reasoning, in the context of the Birch et al. and Crump et al. pain-oriented behavioral criterion 5 for sentience, one of the goals s distinguishing simple reflexive *nociception* from nonreflexive *sentience*. Indeed, as I will discuss in greater depth in later chapters, criteria 4–8 entail pain-related behaviors that would be considered nonreflexive responses, the extent to which they are longer in duration, more flexible, entail more cognition and "trade-offs" between options, and more likely to involve enduring and motivated behaviors.[13]

For instance, for criterion 5, *Motivational trade-offs,* that entails weighing positive versus negative outcomes of an action, they explicitly infer that the behavior is, in this circumstance, nonreflexive in nature: "trade-off behaviour offers relevant evidence, by showing that the nociceptive response goes beyond a reflex and involves centralized, integrative processing" (Crump et al., 2022b, p. 7).

The same applies for their criterion 6, *Flexible self-protection.* Here the animal shows flexible self-protective behavior in a behavioral response that is directed to a specific bodily site of injury and a primary rationale for judging this behavior as sentient is its nonreflexivity:

> Here, we are looking for robust evidence of self-protective behaviours that go beyond reflexes—another plausible evolutionary function of sentience. To meet this criterion, the animal should target its response according to where on the body the noxious stimulus was administered, varying the response as if trying different solutions to the problem. (Crump et al., 2022b, p. 19)

Criterion 7, *Associative learning,* is a much-discussed domain of behavioral evidence of sentience. The simplest, least complex, and most reflexive form of learning is called "nonassociative learning." This is said to occur when an organism's behavior changes toward a stimulus despite that stimulus not having any connection with or association to any other stimulus.

The two major forms of nonassociative learning are *habituation* and *sensitization.* Habituation is said to have occurred when a stimulus that is repeatedly presented to an animal causes a progressive *decrease* in the animal's response. Sensitization is the opposite and is said to occur when there is an *increase* in responding after exposure to a typically noxious stimulus. Nonassociative learning is considered the simplest and most reflexive form of learning.[13]

Associative learning is viewed as a less reflexive form of learning. There are also two major forms of conditional associative learning. In *classical conditioning,* or *Pavlovian conditioning,* the animal learns to associate a conditioned stimulus (CS) such as the sound of a bell with an unconditioned stimulus (UCS) such as the presence of food. The learning is said to occur when then the animal responds to the conditional stimulus in the absence of the unconditional stimulus.

Another type of associative learning is called "operant" or "instrumental conditioning." In operant conditioning, an animal learns to associate their *own behavior* with a reinforcer. While these more sophisticated learning and memory operations are not *directly* related to the "feeling" qualitative aspects of sentience in the same way as are mental images and affects, they are more likely to indicate learned valences and memory of enduring, nonreflexive, and motivated affective states. On this basis, operant conditioning is considered by many including another reasonable *indicator* of sentient feeling.[14]

In their commentary on the Crump et al. criteria, Eva Jablonka and Simona Ginsburg, who have written extensively about the behavioral aspects of associative learning with reference to sentience[15] provided this succinct summary of their view of this relationship:

> As we have suggested elsewhere (Ginsburg and Jablonka 2019), open ended associative learning (in this case of composite predictors of aversive or aversion-ameliorating stimuli or actions)—which requires multimodal discrimination, motivational tradeoffs, instrumental goal-directed conditioning, trace conditioning and second order learning—is a very strong indicator of sentience, accompanied invariably by integrative brain areas that support it. Such learning has been shown to be possible (in humans) only when there is conscious awareness (as noted also in the target article). On-line updating and prioritizing requires the same functional cognitive-affective architecture. (Jablonka and Ginsburg, 2022, pp. 2–3)

Note that their analysis also incorporates the features of motivational trade-offs and goal-directed behaviors, as well as the requisite *neural infrastructure* to support these behavioral features. And note as well that these behaviors are a far cry from any nociceptive-related behaviors that might be considered reflexive.

To meet criterion 8, *Analgesia preference,* the animal when injured learns to locate and self-administer analgesics or anesthetics and is motivated to obtain these compounds over other needs (such as food) when injured. Here, again, this criterion also demonstrates the nonreflexive nature of these proposed behaviors as they include, at a minimum, the motivated choice of pain relief and trade-offs between pain and food.

Finally, Lynne Sneddon et al. proposed that there is an important relationship between evidence for conscious experience of pain and the *complexity* of behavioral responding to noxious stimuli and its nonreflexive nature as important indicators of conscious pain in animals. They pointed out that a number of indices have been offered as evidence for conscious pain experience but the complexity and nonreflexive nature are critical features of these behaviors:

> That together they represent an increasing level of complexity of responses to pain that go beyond simple and acute detection and reflex responses and begin to demonstrate a level of behavioural complexity that would require some form of experience. Pain is a complex multidimensional phenomenon (Rutherford, 2002); therefore, effectively identifying and then assessing the severity of pain may require a multimodal approach. (Sneddon et al., 2014, p. 209)

I will have more to say about the connection between complexity and the emergence of sentience in the next chapter.

Other Nonreflexive Affectively Motivated Behaviors

In addition to the largely pain-related sentient criteria proposed by Birch et al., I will also discuss some other nonreflexive, nonconditioned, motivated, goal-directed, affectively related behaviors. These occur more naturally, in the way that flexible self-protection occurs without specific associative learning constraints. For instance, I will review some naturally occurring escape behaviors that would fall within this category that can serve as additional examples of sentient behaviors.

Summary

In summary, I will consider the presence of isomorphic sensory images and interoceptive-affective awareness subtypes of sentience and that the various neuroanatomical and behavioral features can be used to distinguish reflexive sensing from nonreflexive sentience. In what follows, I will try to show how these subtypes of sentient experiences are novel emergent features of sufficiently neurobiologically complex and variably neurohierarchical central nervous systems.

I hypothesize that as we trace the stages of the emergence of sentience that I propose in the next chapter, while there are differences between these two major subtypes of sentience that I hypothesize, they tend to cooccur within individual organisms at any given stage and that they emerge at the same neurobiological and evolutionary levels. This suggests that these two subtypes likely *coevolved*.

Further evidence for this proposal is that despite the differences between sentient subtypes, there is also a good deal of overlap between the evidence for exteroceptive and interoceptive-affective sentience. For instance, Birch and Crump's criterion 2, *sensory integration,* is a requirement for both exteroceptive sensory and interoceptive-affective sentience. Also, both subtypes have in common unique and advanced neural architectures (*infrastructures*) that are necessary for the emergence for exteroceptive sentience and interoceptive-affective experiences. I discuss this next.

4 The Stages of the Emergence of Sentience: General Principles

Now that the general subtypes of sentience and their criteria are outlined, in this chapter I discuss some general principles for marking the progression from sensing to sentience and how this helps us to distinguish nonsentient (simple sensing) from sentient animals.

As I discuss in this chapter and later on, I propose that basic sensing capabilities are an evolutionarily early emergent feature of all organisms. But sensing does not constitute sentience. Rather, I propose that sentience is a *neurobiological feature of sufficiently evolved and complex brains*, and while it evolved from *living* single-celled organisms, it took as much as 3 billion years to emerge.

Nonetheless it is also my view that it is reasonable and logical to suppose that sentience is an emergent feature of *both* sensing and life (also see chapters 10 and 11). But in order to trace this long road from sensing to sentience, we will need some general neurobiological features that mark this emergence.

Neurobiological Complexity, Evolution, and Emergence

Neurobiological Emergentism (NBE) proposes that, in its simplest and summarized terms, the magnitude or degree of four *objective* neurobiological and evolutionary variables determine the emergence of sentience. These four variables are: (1) the *number* of neurons in the nervous system; (2) the degree of the *specialized functions* of these neurons; (3) the number of *neuro-hierarchical levels*; and (4) the degree of the *interaction* of these levels. These diverse but related variables lead to the emergence of sentience (table 4.1)

So when I refer to either an increase in the *neurobiological complexity* of a nervous system, or more *neurobiologically evolved brains*, I am simply

Table 4.1
Neurobiological variables that mark the progression from sensing to sentience

- Increasing *number of neurons*
- Increasing degree of *specialized neural functions*
- Increasing *number of neurohierarchical levels*
- Increasing degree of the *interaction* between neurohierarchical levels

referring to an increase in the magnitude or degree of these four features; and the increases in these features tend to cooccur in the evolution of nervous systems and the emergence of sentience.

This is, of course, a highly condensed list of the factors that are involved in the neurobiological emergence of sentience that I discuss in chapters 5–7 and these other more fine-tuned features are listed in the tables in these chapters (see tables 5.1, 6.1, and 7.1). But I find that these four general variables can serve as useful shorthand for marking the neurobiological complexity that leads to the emergence of sentience.

Therefore, with reference to the general relationship between complexity and emergence, I use the term "complexity" in a relatively constrained and specific way, that being that with the increase in the aforementioned four objective neurobiological variables, there is also an increase in the *neurobiological complexity* of the nervous system. And with that increase in neurobiological complexity, there is also an increase in novel emergent features that ultimately lead to the emergence of sentience.

This view is at least consistent with some opinions regarding the relationship between complexity and emergence in general. For example, emergent features are often viewed as maximized in *complex systems* with many interacting parts. Thus, the more interacting parts that there are in the system, the greater is the likelihood that novel emergent aggregate system features will be created.[1]

Recall from chapter 3 that Sneddon et al. commented on the relationship between the *complexity* of behavioral responding to noxious stimuli and its *nonreflexive nature* when trying to determine the presence of conscious pain in animals. Note further that a relationship between system complexity in general and emergence has also been noted by many authors.

For instance, James Ladyman and Karoline Wiesner in a comprehensive book on complexity science note that "there is no conception of complexity or complex systems that does not involve emergence" (Lady-man and Wiesner, 2020, p. 73).

Melanie Mitchell, another authority on complexity and systems theory, goes so far as to suggest that a potential definition of a complex system should include the presence of emergent features in self-organizing systems such as living organisms:[2]

> Systems in which organized behavior arises without an internal or external con-troller or leader are sometimes called *self-organizing*. Since simple rules produce complex behavior in hard to predict ways, the macroscopic behavior of such sys-tems is sometimes called *emergent*. Here is an alternative definition of a *complex system*: a system that exhibits nontrivial emergent and self-organizing behaviors. The central question of the sciences of complexity is how this emergent self-organized behavior comes about. (Mitchell, 2009, p. 13)

Many writers who study complex systems theory agree that complexity is one of the key factors in the creation of emergent features. So complex sys-tems and emergence are interrelated; and system complexity and emergence can also be related by the neurobiological complexity and the emergence of sentience. Throughout this book, we will explore how neurobiological com-plexity and emergence is especially magnified in neurohierarchical systems. This is one of the foundations of and sources of support for NBE.

The Neurobiological-Evolutionary Model of the Emergence of Sentience

The model that I present in the following three chapters proposes that there are correspondingly three broadly drawn stages in the emergence of sen-tience (figure 4.1).

Emergent stage 1 (ES1): Single-celled organisms; first appearance 3.5–3.4 billion years ago. At this stage, there are the *lowest levels* of neurobio-logical evolution in the progression from sensing to sentience. Here, the capabilities that are present in single-celled organisms without neurons or nervous systems are best characterized as basic reflexive *sensing*.

Emergent stage 2 (ES2): Neurons, nervous systems, and evolutionarily early brains; first appearance: ~ 570 million years ago. I propose that these are at the *midlevels* of the neurobiological factors that contribute to the emer-gence sentience. I will discuss in the next chapter why I suggest these animals as *presentient* and why they fall roughly between basic sensing (ES1) and sentience (ES3) along the path from sensing to sentience.

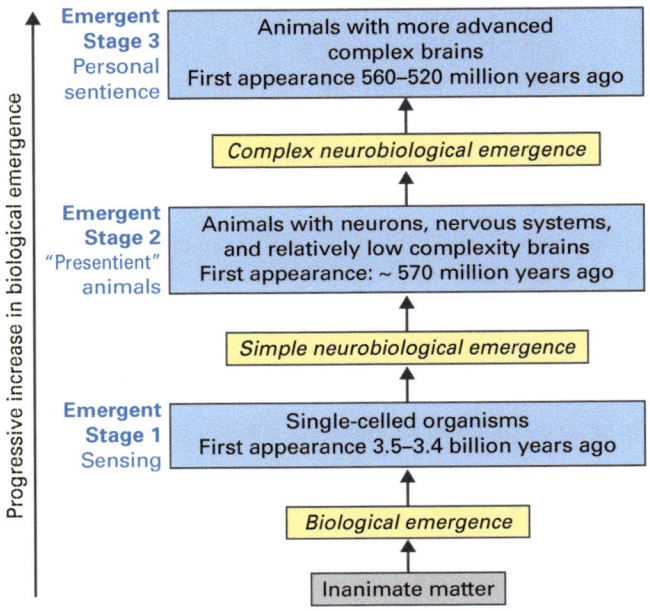

Figure 4.1
The stages in the emergence of sentience. It is proposed that there are roughly three stages—emergent stage 1 (ES1); emergent stage 2 (ES2); and emergent stage 3 (ES3) in the emergence of sentience. Sentience is hypothesized as a naturally occurring emergent feature of life, and that it progresses from sensing to sentience in stages as a result of increasing numbers of neurons, their specialized neural functions, and increasing number of neurohierarchical levels and their interactions (table 4.1). It is hypothesized that the *personal nature of sentience* is a natural result of this progression. This model is based upon Feinberg and Mallatt (2016a, 2016b, 2018, 2019, 2020), with the emphasis now on the role of emergence in the progression of stages.

This level represents a significant advance in the sensing capabilities of an animal when compared to ES1 microorganisms, but falls short of what I hypothesize to be full sentience that emerges in ES3 animals. That said, I also think that the potential sentience of some of these animals is more ambiguous than ES1 (nonsentient) single-celled organisms, and clearly some ES2 animals more than others are further along in the eventual and clear emergence of sentience at ES3.

Emergent stage 3 (ES3): Animals with the collective features of more neurobiologically complex nervous systems and brains; first appearance approximately 560–520 million years ago. At this stage, we witness the higher levels of the emergent factors that contribute to sentience and the clearest evidence of its presence. The evolutionary progression from sensing to sentience makes a substantial leap between ES2 and ES3.

In the next chapter, I first consider ES1 organisms, and in chapters 6 and 7, I go over ES2 and ES3 animals, respectively.

5 Emergent Stage 1: Single-Celled Organisms and the Emergence of Sensing

All living systems, no matter how simple, are replete with examples of basic emergent system features and functions.[1] Even the simplest single-celled organisms that arose billions of years ago possess emergent biological processes that are far more complex than the nonliving substrates of which they are composed.

Implications of Emergent Stage 1 Features

Therefore, the next question is: How do these general features of the emergence of life relate to the emergence of sentience? First, note that ES1 features (table 5.1) entail all the general features of biological emergence that are listed in table 2.1 in that they require multiple parts and multilevel *aggregate functions* of the parts of that system, and the *embodiment* of the cell is essential for the integration of the processes of the parts of the cell that are required for the emergence of novel features. From this, I reiterate, the life of a single-celled organism is an emergent aggregate system feature of the atoms, molecules, proteins, membranes, ribosomes, and so on, and their interactions.

Regarding the eventual emergence of sentience, the most important feature at this level of biological emergence is the capacity of these animals to *sense* the environment, to be responsive to internal homeostatic states, and to show adaptive responsiveness to these stimuli and states. In fact, as summarized by Haswell et al., basic sensing capabilities are characteristic *of all living things*:

All organisms, from single-celled bacteria to multicellular animals and plants, must sense and respond to mechanical force in their external environment (for

Table 5.1
Emergent stage 1: Single-celled organisms such as modern-day prokaryotes *Escherichia coli*, and the eukaryotes *Amoeba* and *Stentor*

First appearance of protocells ~ 3.7 billion years ago; of prokaryotes ~ 3.5–3.4 billion years ago.

Evolutionary and biologically novel structures:
• Macromolecules (proteins, nucleic acids, sugars, lipids), organelles, cells.
• *Embodiment*: semipermeable membrane encloses cell contents to concentrate the chemical reactions and processes to sustain life.

General novel emergent processes:
• Life
• Metabolism, to convert food to energy (ATP) and make new cellular materials; efficient use of energy and of vital molecules slows entropy (energy waste lost as heat).
• Homeostasis: maintaining a constant internal environment in response to changes in the external environment.

Novel but nonsentient emergent processes specifically related to sentience:
• The presence of *embodiment* provides self-organization into a singular aggregate cell so that these simple organisms show most of the general emergent features of life that are enumerated in table 2.1.
• Capacity to sense and be responsive to the environment.

(Adapted from Feinberg and Mallatt 2016a, 2019, 2020)

example, shear force, gravity, touch) and in their internal environment (including osmotic pressure and membrane deformation) for proper growth, development and heath. (Haswell et al., 2011, p. 1356)

So even the simplest one-celled organisms are capable of basic *sensing* mechanisms across numerous modalities. Indeed, these sensing abilities are essential to their survival and only those organisms that sustain themselves and avoid the potentially destructive forces of the external environment can survive and reproduce. It logically follows that basic sensing capabilities must have been present with the emergence of the earliest form of life, the *protocells* that appeared around 3.7.billion years ago.[2] And from these protocells evolved the ancestors of modern-day single-celled organisms.

The present-day single-celled organisms are roughly classified into two main groups. The first to evolve were the *prokaryotes*, a group that includes modern-day bacteria that lack a nucleus or any internal membranes. The first fossils of prokaryotes date back to between 3.5 and 3.4 billion years ago. Later, a second group appeared, the *eukaryotes* that have a nucleus. The oldest generally accepted fossils of eukaryotes date back to about 1.5 billion

years ago.[3] Note that it took approximately 2 billion years for this important transition to occur. These time spans will be of interest when we look
at the emergence of sentience from sensing.

Sensing in Bacteria (Prokaryotes)

We now know a good deal about the sensing capabilities and related behaviors of prokaryotes. One domain of these responses is called "bacterial chemotaxis" that occurs via the biasing of movement of the organism toward
the regions of the environment that contain higher concentrations of ben-
eficial, or lower concentrations of potentially harmful chemicals.[4]

For instance, let's consider the sensing and responding chemosensory capabilities of the single-celled bacterium *Escherichia coli* (figure 5.1).
Despite its relatively simple biological anatomy and lack of neurons or a
nervous system, *E.coli* has sensing mechanisms that enable it to adaptively
respond to an array of environmental stimuli including the concentrations
of nutrients and toxins, oxygen levels, pH, osmolarity, temperature, and
the intensity and wavelength of light.[5] The signaling pathways that are
involved in representative bacteria including *E. coli* have been extensively
investigated, and the biochemical details for many these chemosensory
pathways in prokaryotes such as *E. coli* are known[6] (figure 5.2).

How *E. coli* navigates in its generally watery environment is also relatively simple but nonetheless remarkable (figure 5.3). For its direction of
locomotion, *E. coli* utilizes a system of flagella that are often described as
a sort of biological "rotary motor." This involves a system of 5–10 flagella
that are randomly distributed on its cell surface. The response to environmental stimuli is directed by coordinated alterations in the rotational direction of the flagella. When environmental conditions are favorable, all of the
flagellar motors collectively rotate in a counterclockwise fashion that causes
the bacterium to swim smoothly in a forward direction. But when moving
toward less favorable or aversive conditions, the motors switch to a clockwise direction that causes the individual flagella to rotate in different and
more independent directions causing the *E. coli* to "tumble" and alter its
course.

This is called a "run-and-tumble" motion and it is a *random* walk in
which the animal does not actually "choose" the direction that it swimming. In other words, the "sensing" behaviors of *E. coli* are reflexive as
defined by the criteria outlined in chapter 3. But by the coordinated action

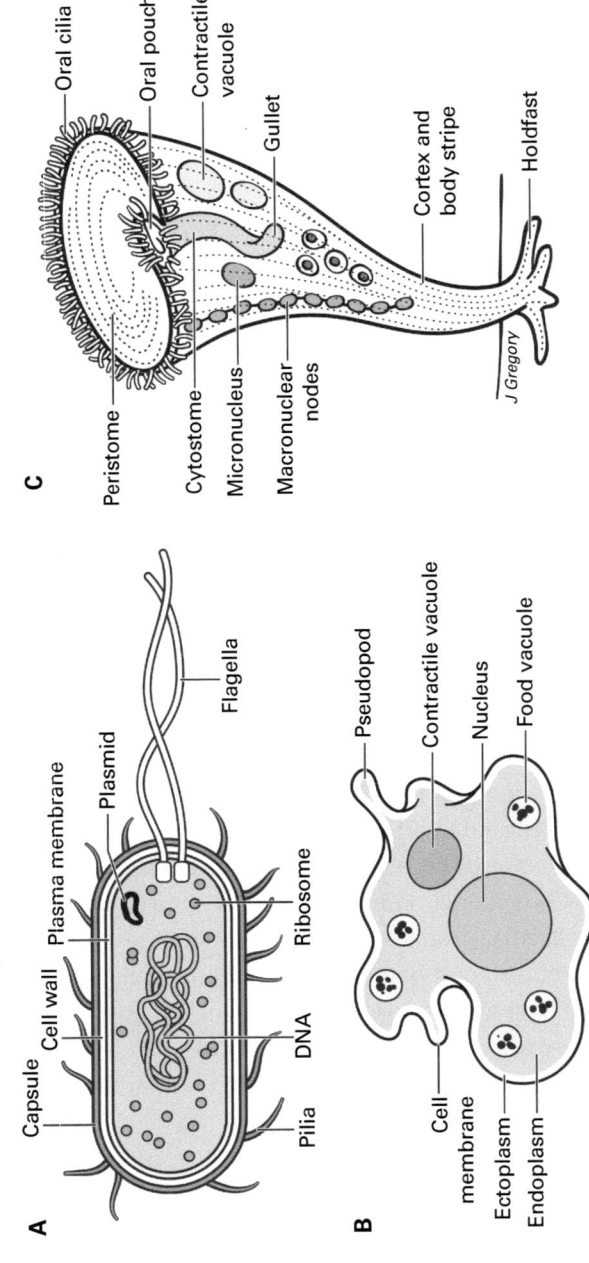

Figure 5.1

Emergent stage 1 single-celled organisms: (A) *Escherichia coli*; (B) amoeba; (C) *Stentor roeselii*.

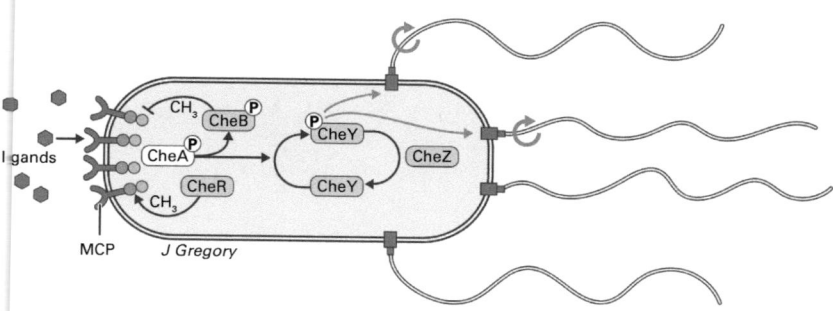

Figure 5.2
Chemotaxis pathways in the prokaryote *E. coli,* which has complex sensing systems that enable it to adaptively respond to an array of environmental stimuli including the concentrations of nutrients and toxins. In this schematic illustration, the organism responds to extracellular ligands (hexagons) that activate transmembrane chemoreceptors (MCPs) and response regulators (CheB, and CheY) in a complex coordination with phosphatase CheZ that determines the probability of "tumble or runs." (Figure based on Micali and Endres, 2016.)

of its sensing structures and their hard-wired connections to the motor apparatus, the animal will probabilistically end up in more favorable conditions (figure 5.3).[7] So we see how this sensing behavior is adaptive, just like photosynthesis or metabolism processes are adaptive, but it is nonetheless reflexive.

Is There "Something It Is Like" to Be a Single-Celled Organism?

The argument has been made by some that single-celled organisms are indeed sentient.[8] One eukaryotic organism—the ciliate *Stentor roeselii*—has become particularly famous in part as a result of this debate (figure 5.1).

S. roeselii is a member of the genus *Stentor,* which are referred to as "protozoans," a relatively nontechnical term for a group of single-celled eukaryotes that also includes amoebas and paramecium. They are found in watery environments, are between 500 and 1,200 micrometers in length, and have a trumpet-shaped front that resembles a funnel. It is a *ciliate* and is covered by hair-like cilia on its body's surface that aid the organism in feeding and swimming through the water.[9]

Although we don't have much information about the sensing capabilities of *S. roeselii* specifically, especially when compared with what we know

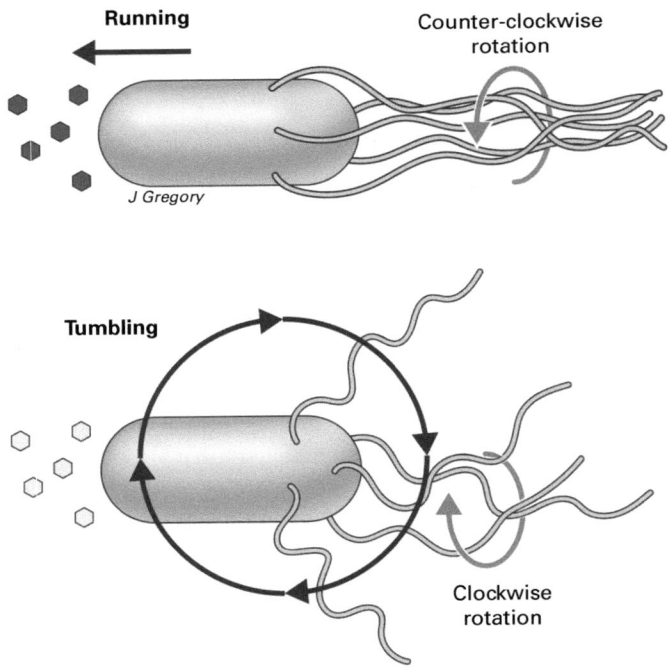

Figure 5.3
The "run-and-tumble" motion of *E. coli*. In favorable environmental conditions, the flagellar motors collectively rotate in a counterclockwise fashion, causing *E. coli* to "run" in a forward direction. When moving toward aversive conditions, the motors switch to a clockwise direction, causing the flagella to rotate in more independent directions that cause *E. coli* to "tumble," thus altering its course. (Adapted from Egbert et al. 2010.)

about these abilities in *E. coli*, we do know that, as a group, species of *Stentor* and other ciliates have a full array of sensing capabilities, including chemosensing and mechanosensing abilities, that are typical of single-celled organisms in general.[10]

Herbert Spencer Jennings, *S. roeselii*, and the Question of Stentor Avoidance/Escape Behavior

The oft-told story of how *S. roeselii* became famous goes back to 1902 when zoologist Herbert Spencer Jennings became interested in its behaviors. Most significantly, Jennings reported on the *Stentor's* escape behavior when it was exposed to a noxious stimulus—a red dye called "carmine power"—that Jennings delivered by pipette in front of the ciliates mouth.

S. roeselii can be free swimming or sessile, moored in place by its "foot." The sequence of interest begins with *S. roeselii* in its sessile position, which typically entails it being nested within a "tube" that it builds out of external debris and its own mucus.

When Jennings introduced the dye, he reported what he considered to be a roughly sequenced hierarchy of avoidance-escape responding.[11] First, the animal "bends away" from the dye, this action minimizing its exposure to the dye particles. If that doesn't work, the organism changes the direction of the beating of its cilia, with this action resulting in altering the direction of water currents and reducing the number of particles in contact with its mouth. If both of these actions failed, Jennings reported that the *S. roeselii* would contract its body into its tube. And finally, if these reactions weren't sufficient, then it would detach itself from its tube and swim away (figure 5.4).

In Jennings's view, these behaviors were not simply occurring randomly or reflexively. Rather, he interpreted this behavior as *Stentor* intentionally trying to escape the dye:

> There are variations in the details of the reaction series under different conditions. Sometimes, one step or another is omitted, or the order of the different steps is varied. But it remains true that under conditions which gradually interfere with the normal activities of the organism, the behavior consists in 'trying' successively different reactions, till one is found that affords relief. (Jennings, 1906, p. 177)

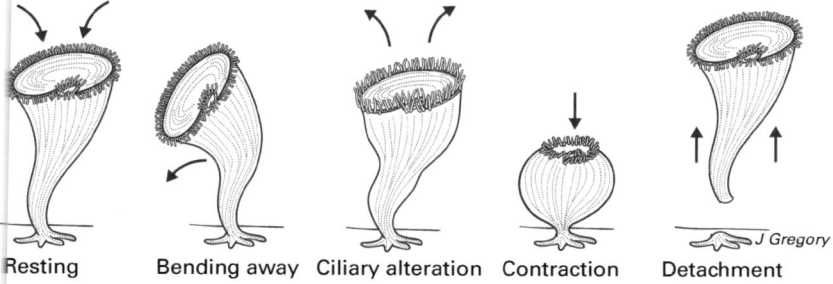

Resting   Bending away   Ciliary alteration   Contraction   Detachment

Figure 5.4
A sketch of avoidance hierarchy in *S. roeselii*. See text for details. (Adapted from Dexter et al., 2019.)

While some in the scientific community found Jennings' findings of interest, a paper published many years later, in 1967 and in the wake of the popularity of behaviorism, reported that these investigators were unable to replicate Jennings' findings.[12] However, the experiments in that paper were performed with a different species of *Stentor,* which certainly could have affected the findings.

Then, subsequently, in an experiment reported in 2019 paper by Dexter et al., this group again tried to replicate Jennings' findings[13] but this time with *S. roeselii,* the same species of *Stentor* used by Jennings. While they could not replicate the avoidance behavior using modern-day carmine powder, they found that pulses of polystyrene beads in an aqueous suspension with sodium azide (NaN_3) were effective in eliciting the avoidance response.

The results of this experiment are complicated. This team was indeed able to replicate some of Jennings' findings. For instance, they found that the avoidant response they recorded replicated the four avoidant behaviors that Jennings described—that is bending away, ciliary alteration, contraction, and detachment from the tube. And they concluded by using a sophisticated statistical analysis that indeed a hierarchy of behaviors was to some extent present.

But there are some caveats regarding the strength of the behavioral hierarchy. For instance, as Jennings observed, these authors also reported that there was "substantial heterogeneity" in the sequence of escape behaviors and that they found "few instances of the full hierarchy, but many partial instances with varying orders of occurrence of individual behaviors."[14] And they also found that the *Stentor's* "choice" between contraction and detachment was consistent with a "fair coin toss." So rather than there being a "logical" and nonreflexive pattern of behavioral responses, the "hierarchy" of responding appears to be more of a statistical probability of the occurrence of certain adaptive behaviors rather than a strict nonreflexive behavioral hierarchy.

The Question of the Unicellular Consciousness

As aforementioned, there are some proponents of the view that the behaviors of these unicellular organisms down to prokaryotes such as *E. coli* are sentient. One of the strongest advocates for this claim is cognitive psychologist

Arthur Reber who proposed what he called the "Cellular Basis of Consciousness (CBC)" model in which the behaviors of single-celled microorganisms—one of his favorites also appears to be *Stentor*—represent not just sensing abilities but sentience (consciousness). In the case of a bacterium:

> A bacterium does not just sense that there is a gradient of sugar molecules in the surround, it perceives its positive valence and actively moves in the direction of the higher concentrations. It does not simply detect that an acidic molecule is touching its surface membrane, it interprets it as an aversive stimulus and retreats from it in a distinctly measured way. (Reber, 2019, p. 139)

Notice the similarities between Reber's and Jennings' points of view. In Reber's case, the bacterium is not simply detecting or sensing but "perceiving" a stimulus; and the bacterium "interprets" the aversive nature of the stimulus, similar to Jennings' view that the *Stentor* is "trying" successively different reactions, "till one is found that affords relief."

This begs the question of how do the behaviors of ES1 organisms match up with the criteria for sentience that I proposed in chapter 3. First, there is no question that these simple organisms are able to *sense*; but they are not neuroanatomically capable of creating neurohierarchical sensory maps or mental images. Second, although they are capable of responding to positive and negative stimuli in a valenced-appropriate and adaptive manner, they do so without any of the neuroanatomical infrastructure that is required for interoceptive-affective feeling. Finally, the sensing behaviors of single-celled organisms are fully explained by nonsentient and reflexive biochemical mechanisms.[15]

So, my final conclusion must be that these behaviors—no matter how complex and sentient they *appear*—are nonsentient and can be fully explained by standard biological mechanisms in the same way as any other biologically emergent process is explained.

I will further examine the claim that all life including single-celled organisms are conscious in the context of emergence in chapter 9.

Summary

I therefore conclude that *sensing* in single-celled organisms has a long way to go—indeed some billions of years of evolution—before the degree of emergence required for sentience is reached. I will describe in the ensuing chapters how and why sentient animals have a veritable explosion of novel neural, neurobiological, and emergent features that are required for the creation of sentience and that these are not present in ES1 organisms.

So while I applaud the connection between life and sentience that proponents of unicellular sentience endorse, I will propose that there is an uninterrupted roughly step-wise progression between the *sensing* of bacterium and the *sentience* of some animals that occurred over billions of years, but one without any fundamental gaps. In the next chapter, we move on to emergent stage 2.

6 Emergent Stage 2: Neurons, Nervous Systems, and Evolutionarily Early Brains—"Presentient" Animals

After the evolutionary appearance of the eukaryotes, the first multicellular animals made their appearance around 700 to 600 million years ago. There is evidence that these early multicellular animals were most likely sponges.[1] Modern sponges have some cellular differentiation with just four different cell types, they are immobile, and they have no brain or nervous system. Therefore, the next important evolutionary step for our understanding of sentience would have been the appearance of animals with more cell and tissue types, including *neurons* and *nervous systems*.

The first nervous systems were likely structures called "nerve nets" that existed in early jellyfish-like animals and simple marine worms. Unlike in animals with centralized nervous systems and brains, nerve net are distributed over the body and largely decentralized, but their appearance was an important innovation because this gave them significant adaptive advantages for obtaining food, defense, and mating.

The cnidarians, a phylum of animals that includes modern-day jellyfish, retain these early nervous systems.[2] The nervous system of these species vary. In the class that are often referred to as "true jellyfish," the nervous system includes among other structures a diffuse nerve net and nerve rings (figure 6.1). The most elaborated sensory organs are called "rhopalia." These structures are multiple and contain pigmented photosensitive *ocelli*—light sensitive "eyes"—and *statocysts* that are responsive to gravity and allow for the animal's spatial orientation in the water. The diffuse nerve net of these animals also has simple sensory receptors in the form of nerve endings for detecting touch.[3]

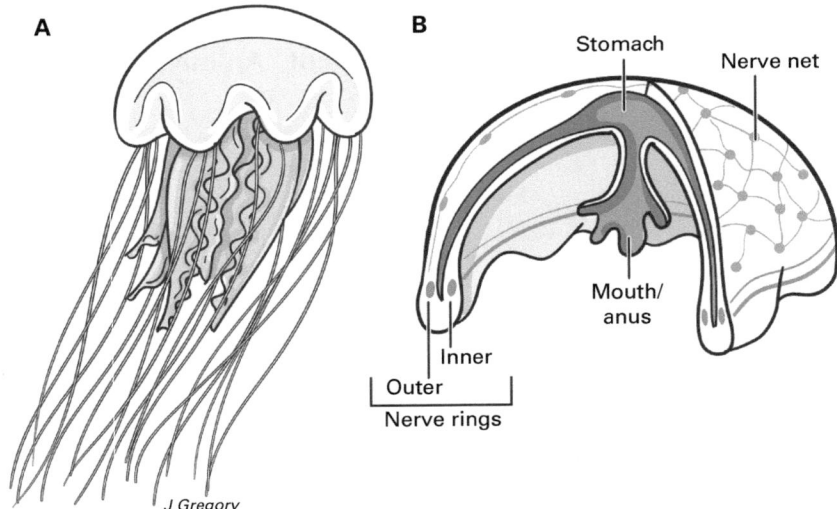

Figure 6.1
Jellyfish (A) and its nervous system (B). The diffuse nerve net of these animals also has simpler sensory receptors in the form of nerve endings for detecting touch.

The Development of Central Nervous Systems: Bilaterians

The first worms had evolved by 570 million years ago, as early bilaterians, which are the animals with matching right and left sides. Over the next 30–50 million years, these early worms gave rise to today's bilaterian animals, including invertebrate groups that lacked a spine and the vertebrates with a spine.[4]

In one proposed scenario, in many of these descendant lineages, some nerve net structures were enlarged and centralized in the head region that received sensory information first as the animal advanced in the environment. There were also neural enlargements in longitudinal nerve cords that would carry sensory information and motor commands to and from the brain. These were the first brains and nerve cords and the beginnings of central nervous systems.[5] Many living invertebrates reflect this stage (e.g., roundworms, earthworms, flatworms, sea slugs, and the fish-like cousin of vertebrates called "amphioxus" that are discussed later).[6]

While these ES2 invertebrates have relatively simple nervous systems and brains, they do possess many novel emergent features that will progress toward the emergence of sentience (table 6.1).

Table 6.1

Emergent stage 2: Animals with neurons, nervous systems, and evolutionarily early brains ("presentient" animals).

First appearance: ~ 570 million years ago.

Organisms at this stage: most invertebrate animals; for example, most worms.

Novel neural structures and processes

- *Multicellular animal body with diverse cell types* including neurons, neural reflex arcs, sensory receptors, and motor effectors (muscles, glands).
- *Neurons* are now able to transmit much fast interneural signals.
- *Nerve nets, then a consolidation into central and peripheral nervous systems*; some of the animals have a simple brain with movement-patterning circuits; the sensory receptors are mechano-, chemo-, and photoreceptor cells.
- *Rapid transmission of signals* enables the animals to control the actions of multi-cellular body in response to sensory stimuli.
- *Connectivity*: reflex arcs and neuron networks coordinate all the parts of a multi-cellular body.

Novel emergent processes related to sentience

- Significant increases in neurohierarchical exteroceptive processing and approach-avoidance adaptive behaviors
- Basic motor programs and central pattern generators for rhythmic locomotion, feeding, and other stereotyped movements.
- Behaviors become less reflexive and more neurobiologically complex when compared to Emergent level 1 organisms.

(Adapted from Feinberg and Mallatt 2016a, 2019, 2020)

In the last chapter, it was relatively easy to show that ES1 organisms have sensing capabilities that are reflexive and hence not sentient. But in ES2 animals with nervous systems, especially those with more evolved brains, there is such a giant leap forward in neurobiological features that judging the presence of sentience becomes more difficult. I ultimately chose the term "presentient" to describe this stage because there is a broad range of ES2 capabilities from animals that are lowest in the degree of emergent features and therefore closest to ES1 organisms and those that are highest and closer to ES3 sentient animals.

Some Emergent Stage 2 Species

Here, I discuss four species of animals that are within the ES2 level that come from different phylogenetic lines and different degrees of neurobiological emergence: the nematode worm *Caenorhabditis elegans;* the cephalochordate *amphioxus,* an animal on the phylogenetic line that leads to sentient vertebrates; and the gastropod mollusks *Aplysia and Pleurobranchaea* (sea slugs) that are on the phylogenetic line leading to sentient coleoids (octopus, squid).

The primary focus of this chapter will be where these species fall in the evolution of sensing to sentience. So, in addition to their basic sensing capabilities, when compared to ES1 animals, they show more complicated capacities to discriminate positive (for instance food) or negative (noxious) nature of an external stimulus and adaptively respond with approach (positive stimuli) or avoidance or escape (negative stimuli) behaviors.

While there are still some important aspects of sentience that are absent in these ES2 animals, there is also a clear progressive increase in the factors that lead to the emergence of sentience at ES3.

Caenorhabditis elegans

The ES2 animal *Caenorhabditis elegans* is a nematode worm, a member of the phylum Nematoda. It is an invertebrate, meaning it has no spine. It is quite tiny—only about 1 mm long and visibly transparent. It lives in the soil and actually eats the microorganisms such as the bacteria that are ES1 organisms we spoke about earlier. It's good to be an ES2 animal!

Nematodes comprise one of the largest phyla in the animal kingdom, both in terms of individual numbers and species diversity. According to William Schafer, while 20,000–30,000 nematode species have been described, the true number of individual species may actually be between 100,000 and 10 million.[7] Most of *C. elegans* are hermaphrodites or males.

The animal's general body schema features an exterior cylindrical exoskeleton called a "cuticle" that gives the animal its shape and protects it from the environment but is also sufficiently flexible to navigate in the soil.

Nervous System of Caenorhabditis elegans

The hermaphrodite version of *C. elegans* is composed of just 959 cells and its nervous system is comprised of only 302 neurons.[8] Despite this apparent simplicity in pure number of neurons, the nervous system of *C. elegans* is remarkably neurobiologically complex when compared to ES1 microorganisms.

Schafer provides the following basic description of its neural anatomy.[9] Located at the core of its central nervous system is a *circumoral brain* or *nerve ring* that is comprised of axons and dendrites whose cell bodies are located anterior and posterior to the nerve ring. These are referred to as "ganglia."

There is a primary ventral nerve cord that runs longitudinally from the worm's head to its tail along its ventral midline whose neurons are reciprocally connected to neurons in the nerve ring. There are other nerve cords, the most prominent among them a dorsal nerve cord that runs down the dorsal midline and there are also numerous circumferential commissures (figure 6.2A).

Sensory systems and neurohierarchical features   The sophistication of sensory capabilities of *C. elegans* is well beyond that seen in ES1 single-celled microorganisms. In addition to chemosensory and mechanosensory organs located around the mouth, the main sensory organ of *C. elegans*

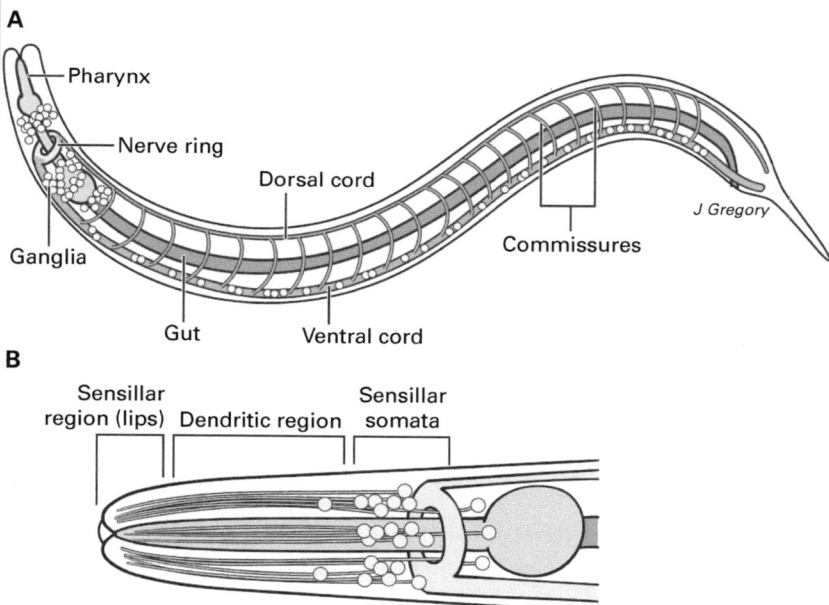

Figure 6.2
C. elegans neuroanatomy and sensilla. (A) shows the general schema of the *C. elegans* nervous system. (B) illustrates the *amphid sensilla* that are the animal's primary chemoreceptive olfactory and taste organs.

consists of two bilaterally symmetric *amphid sensilla* adjacent to the mouth (figure 6.2B). The amphids are the animal's primary chemoreceptive olfactory and taste organs. They also contain somatosensory neurons that are responsive to touch and temperature. These receptors are also located along the surface of the body. Each amphid has twelve sensory neurons that are connected via dendrites to the related sense organs and via axons to the central nerve ring.[10] The animal also has interoceptive and nociceptor sensitivity to various noxious stimuli including thermal and mechanical stimuli.[11]

Perhaps even more remarkable than these diverse sensory capabilities are *C. elegans'* simple, yet to some degree, neurohierarchical sensory integration that features an array of reflex arcs with several orders of neurons. For instance, in a remarkable study by Brittin and colleagues,[12] they reported a multilayered, multiscale, and modular level of organization that they suggest is simpler yet similar to what is found in more neurobiologically complex brains. They conclude that even in this worm's seemingly simple brain, they find what they propose is a *modular network* architecture that entails sensory *computation, integration, sensorimotor convergence*, and even *brain-wide coordination*:

> By characterizing the spatial embedding of its connectome, we sought insight into the structures that could support a hierarchical, modular and nested architecture in the C. elegans brain. Previous analyses of the C. elegans connectome identified a common feed-forward loop motif among triplets of neurons. Our brain map recasts this local motif as an architectural motif, reminiscent of layered cortical architectures and their artificial analogue, residual networks. Such a 'connectionist' description of a biological brain provides a promising methodology for identifying parallel and distributed circuits. (Brittin et al., 2021, p. 109)

Brittin et al. even suggest some similarity with other invertebrates and even vertebrates: "The C. elegans brain map and its nested architecture might suggest a much closer analogy between the C. elegans neuropil and the coordination between the nano- and macroconnectomes of other invertebrates and even vertebrates" (Brittin et al., 2021, p. 110).

This central idea is also voiced by Schafer who notes that although the brains of worms and mammals are clearly different in many respects, there are also organizational patterns that are similar: "Thus, the macroscopic organization of the C. elegans nervous system shows scale-invariant conservation with the human brain over many orders of magnitude of anatomical complexity" (Schafer, 2016, p. R959).

Presentient approach-avoidance behaviors of *Caenorhabditis elegans* There are some remarkable examples of the increasingly sophisticated sensing and responding capabilities of the *C. elegans* nervous system that combine neurohierarchical sensory organization with more differentiated approach-avoidance behaviors. These represent a significant advance when compared to the simpler sensing and responding capabilities that are shown by ES1 single-celled organisms.

For instance, Metaxakis and colleagues[13] describe how *C. elegans* is capable of integrated and neurohierarchical processing of positive and negative sensory signals. For chemosensory processing (e.g., odor sensation), there is an array of differentiated sensory neurons that are responsive to water soluble attractants and repellents. Then, via another complicated array of downstream interneurons, the worm develops a motor response, such as backward (avoidance) or forward (approach) movement (figure 6.3).

Further, they explain how many of the sensory neurons are polymodal (responsive to more than one sense modality; figure 6.3). This allows the animal to cross-modulate and integrate information from multiple sensory modalities via crosstalk between primary and secondary neurons and interneurons and even integrate this information with the animal's interoceptive state.[14]

There will be more to say about various emergent levels of organization in following chapters. But here, I just want to emphasize that these sensory properties clearly meet the criteria for *novel emergent features* (table 6.1).

Amphioxus: On the Vertebrate Phylogenetic Line

The next ES2 animal that we consider is amphioxus (*Branchiostoma*). Amphioxus, also known as a lancelet, is a fish-shaped marine animal about 4–6 cm long. The adult animals most commonly live in warm ocean environments where they burrow tail first into the sand. It is a member of a group of invertebrate animals called "protochordates," a group that is comprised of two invertebrate subphyla—the cephalochordates that includes amphioxus and the urochordates, also known as tunicates, that includes animals such as the sea squirt.

Protochordates do not have a spinal cord but at some point in their life, they have a structure called a "notochord" and a hollow *nerve cord* that runs along the dorsal surface of the animal. In vertebrates (such as fish, amphibians, reptiles, birds, and mammals), these structures will become the spine

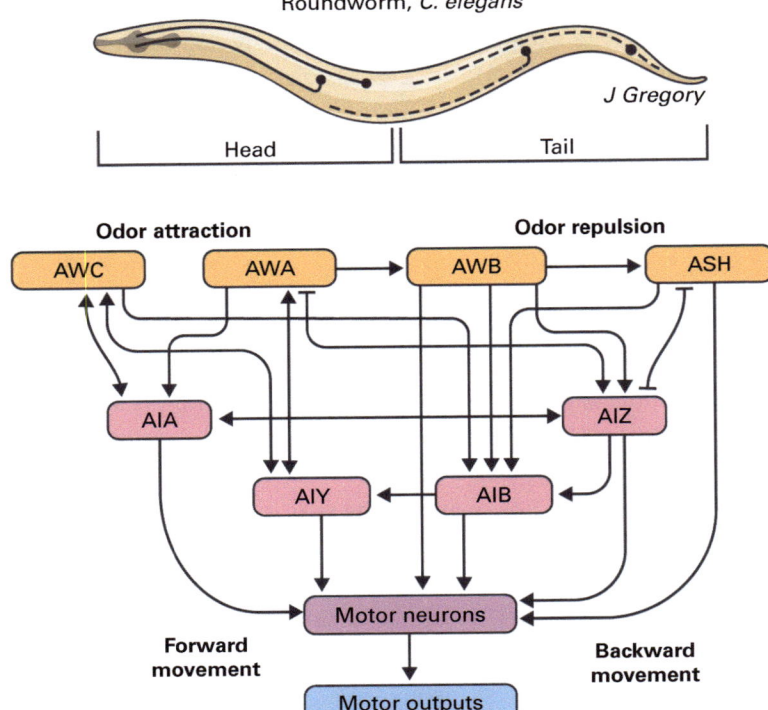

Figure 6.3

C. elegans neural sensory processing of approach and avoidance to odor stimuli. For odor sensation, there is an array of differentiated sensory neurons (orange) that are responsive to water soluble attractants and repellents. Then via another array of downstream interneurons (pink) there are connections to motor neurons (purple) that enable an adaptive motor response (blue) such as backward (avoidance) or forward (approach) movement. The arrows signify chemical synapses and bars electrical synapses (gap junctions). If there are both arrows and bars, that means that these neurons are connected by both chemical and electrical synapses. (Based on Metaxakis et al. 2018.)

and central nervous system, respectively. In amphioxus, the notochord and nerve cord persist their entire lives.[15]

Nervous system of amphioxus The adult brain of amphioxus, although much simpler than any vertebrate brain, still could have upward of 20,000 neurons; while for the larval stages, the number is likely much lower and closer to 300–500 neurons. Numerous studies indicate that amphioxus is the closest living "proxy" for the ancestral chordate (animals with spinal cords) condition, and it has several brain structures that are likely progenitors or homologs of telencephalic, diencephalic, midbrain, and hindbrain structures that are found in later evolving vertebrates. Larval amphioxus also has an unpaired frontal eye in the midline that is the homologue of vertebrates' paired eyes (figure 6.4).[16]

Sensory systems and neurohierarchical features of amphioxus Evolutionary biologist Thurston Lacalli has done extensive studies on the amphioxus nervous system that I summarize here. The body surface of amphioxus is supplied with an assortment of epithelial sensory cells, which typically develop from scattered precursors. He estimates (personal communication) that from the number of morphologically distinguishable sensory cells in the epidermis, amphioxus probably has more sensory cell types than *C. elegans*, including olfactory ones.[17]

Amphioxus has four types of photoreceptors. One type creates a *frontal eye* but Lacalli notes that it is too small and simple to be "image-forming."

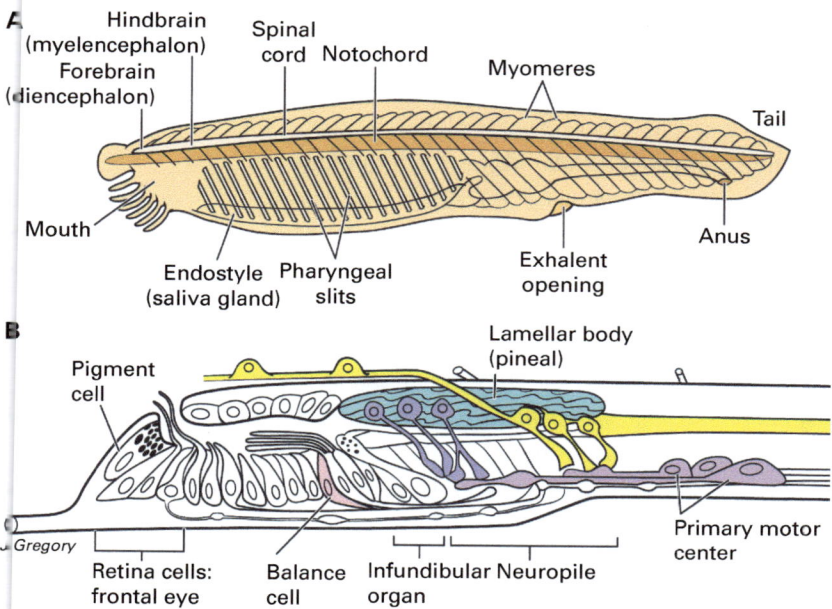

Figure 6.4

Amphioxus general neuroanatomy (A) and the neural anatomy of its escape responding (B). See text for details. (Adapted from Lacalli 2021.)

Nonetheless, it is clear that amphioxus is capable of complex adaptive responding that relies upon relatively simple neurohierarchically integrated sensing capabilities and responses.[18]

Presentient approach-avoidance behaviors of amphioxus   An excellent model for advancing approach-avoidance responding in amphioxus is the escape behavior in young amphioxus as described by Lacalli. What follows is a simplified summary of his analysis.[19]

This model features a critical integrative brain zone that is known as the *post-infundibular neuropile*. This structure receives multiple inputs from the frontal eye as well as an assortment of other types of sensory neurons, from the lamellar body (the analog of the vertebrate pineal gland), and from epithelial sensory cells whose axons enter the nerve cord via the paired rostral and anterodorsal nerves. The animal's escape behaviors are induced most easily by mechanical stimulation of the rostrum, and sensory fibers originating there converge on the *primary synaptic zone*, and along with input from the frontal eye, regulate the animal's escape response (figure 6.4B).

Here, the same points that I raised with reference to nematodes also apply to amphioxus. That is, compared to the simple reflexive responding at the ES1 level, we also witness with amphioxus a more neurobiologically complex, somewhat neurohierarchical adaptive responding to a noxious stimulus.

Sea Slugs (*Aplysia, Pleurobranchaea*): On the Coleoids
(Octopus, Squid) Phylogenetic Tree

The mollusks are a large group of soft-bodied invertebrates that is comprised of several main subgroups. The bivalves are the simplest and include scallops, clams, and oysters. Next in terms of neurobiological complexity are the *gastropod mollusks,* a group that includes snails and several species of slugs including *Aplysia* and *Pleurobranchaea;* the most sophisticated mollusks are the *cephalopod mollusks* such as squid, the nautilus, cuttlefish, and the octopus. The octopus is an ES3 animal that I will discuss in chapter 7.

The largest group among the mollusks are the gastropods that comprise eighty percent of all mollusk species.[20] *Aplysia* (also known as "sea hares") are a genus of gastropod mollusks of which there are thirty-seven identified species. These species can vary greatly in size from several centimeters in length to very large species that measure over 60 cm. *Aplysia* are best known in neuroscience from the work of Eric Kandel who shared the Nobel Prize for his work on learning and memory especially in *Aplysia californica*.[21]

Nervous system of *Aplysia* It is estimated that there are about 10,000 neurons in the central nervous system of *Aplysia*. While this overall number may be more than some other mollusk species, it is far fewer than the cephalopods that we consider in the next chapter. Its central nervous system is organized into nine major ganglia: paired buccal (BG), cerebral (CG), pleural (Pl), and pedal (Pe) ganglia, and a single abdominal ganglion (AG) that are connected via commissures and structures called "connectives" (figure 6.5).

The ganglia—depending upon where they are positioned in the animal's nervous system—have some specialized and regional functions regarding the sensing and behavioral control of the animal. For instance, the caudally located abdominal ganglion controls basic physiological functions including the animal's heartbeat and aspects of its breathing while the most rostrally positioned buccal ganglia is located at the head and controls the feeding apparatus.

The sensory pathways of gastropod mollusks consist of chains of just one to three neurons with neural hierarchies that are much shorter than the ES3 cephalopods that are further along in the evolutionary line. The cerebral ganglion that controls the head and lips does receive input from all the other ganglia and therefore could provide some integrated centralized hierarchical control for the whole animal.[22]

Sensory systems and neurohierarchical features of *Aplysia* *Aplysia* have small, black, pinpoint paired eyes that are located at the head of the animal but are too small to be image-forming. In addition, the animal's auditory capabilities are also limited. However, the animal compensates for this with

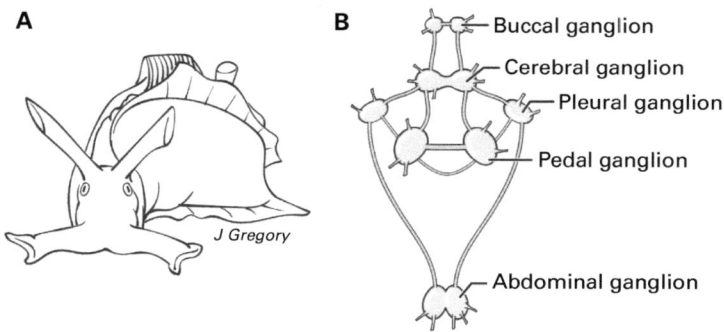

A

B
— Buccal ganglion
— Cerebral ganglion
— Pleural ganglion
— Pedal ganglion
— Abdominal ganglion

J Gregory

Figure 6.5
Aplysia californica (A) and its nervous system (B). See text for details.

its chemosensory and tactile capabilities that enable it to perform various behaviors such as the finding food and mating. Olfaction is particularly important as a distance sensor and the posterior tentacle of *Aplysia*—the rhinophore—is its primary olfactory organ. There is also some peripheral nervous system somatotopic encoding of external stimuli in the "oral veil," a structure that includes bilateral tentacle ganglia and primary receptors, but central neurons for somatotopic maps have not been found.[23]

Approach-avoidance behaviors of sea slugs In an excellent model that demonstrates the increasingly sophisticated approach-avoidance behaviors in *Aplysia*, Gillette and Brown[24] propose a system wherein incoming extero-sensory signals are assigned positive or negative valences in an "integrator circuit for incentives" in the cerebral ganglion (figure 6.5B), which then connect with nearby premotor circuits called "central pattern generators (CPGs)." These communicate with the hierarchically "downstream" pedal ganglia that ultimately control the slug's feeding (approach) or avoidance locomotor movements (figure 6.6).

Are These Emergent Stage 2 Animals Sensing or Sentient?

C. elegans First, with reference to their sensory abilities, while *C. elegans* do have diverse sensory capabilities across multiple modalities, these animals have no integrated and specialized sensory receptors such as image-forming eyes or an auditory system, so they lack the elaborate distance senses such as vision or hearing that would allow them to form mental images of sights or sounds. Put another way, they do not have the neuroanatomy that I and others propose are necessary for exteroceptive mental images. So despite *C. elegans* having some basic neurohierarchical pathways, their brains are not anatomically capable of creating mental images that are universal in ES3 sentient animals.

Second, while *C. elegans* can detect and appropriately respond to noxious stimuli, this is a feature of even single-celled ES1 bacteria, organisms that are not sentient. And while *C. elegans* has nociceptors and is responsive to opioid receptor agonists such as morphine,[25] their nociceptors are relatively sparse[26] and they lack the multisynaptic central neuroanatomical infrastructure that we will see in the next chapter is clearly present in ES3 animals that possess all the features of sentience.[27] So the anatomical case for *C. elegans* having exteroceptive, interoceptive, or affective sentience is not entirely absent but it is not compelling.

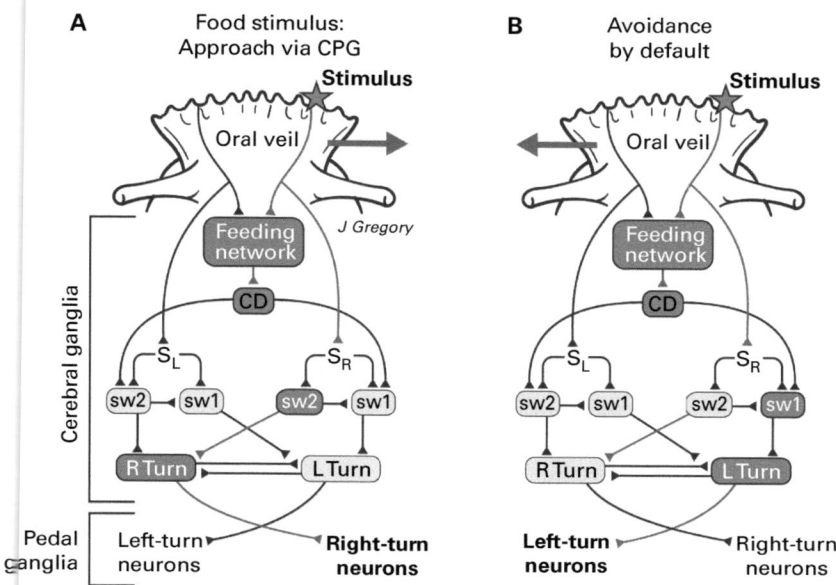

Figure 6.6
Approach and avoidance networks in the sea slug *Pleurobranchaea californica*. The valence system of the sea slug as proposed by Gillette and Brown (2015). Incoming exterosensory signals are assigned positive or negative valences in an "integrator circuit for incentives," in the cerebral ganglion that communicate with nearby premotor circuits called "central pattern generators (CPGs)." The CPGs, in turn, connect with the hierarchically "lower" pedal ganglia that control the slug's feeding or approach (A) or avoidance (B) locomotor movements. (Figure adapted from Gillette and Brown, 2015.) CD = corollary discharge neurons; Sw1 and Sw2 = switch neurons; SL and SR = sensory pathways left and right.

Third, are their behaviors sentient? Barron and Klein presented a good argument against nematode sentience on the basis of the relatively reflexive nature of their behaviors. They argue that while nematodes have a centralized nervous system and can process and adaptively respond to an array of stimuli, their behaviors are not sufficiently goal-directed or motivated (as I mentioned per Berridge in chapter 3) to be clearly sentient, but rather are based upon current stimuli or present interoceptive state. They point out that nematodes don't actively hunt for food beyond their immediate environment and thus their behaviors aren't clearly goal directed as would be required in nonreflexive behaviors and in contrast to sentient mammals and insects. As they put it:

Hence, in mammals and insects homeostatic drives direct behavior to where resources are expected to be, even if they are not currently there. We argue the difference between this behavior and nematode search behavior arises because nematode behavior is organized by reference to their primary sensory input, whereas rodent and insect behavior is organized in response to an integrated and spatial simulation of their environment. Nematodes do possess forms of memory that can change how they react to stimuli but there is no evidence this memory has a spatial component or contributes to a structured model of their environment. Consequently, even though nematodes have a centralized nervous system and memory, they lack the egocentric modeling of the environment that is required for subjective experience. (Barron and Klein, 2016, p. 4905)

In summary, based upon anatomical and behavioral evidence, while *C. elegans* is a clear emergent advance over ES1 unicellular organisms, its features with regard to sentience are best characterized as presentient, especially in comparison to ES3 animals.

Amphioxus

Despite being on the evolutionary line to sentient vertebrates, the sensory neurons of amphioxus are not derived from neural crest or placodes as is the case with vertebrates, so the sensory systems in amphioxus are not homologous in the two animal groups. So while amphioxus has photoreceptors, touch receptors, chemoreceptors on its body surface, and possibly some olfaction, amphioxus lacks image-forming eyes, vision, and hearing so the isomorphic maps that are required to create isomorphic mental images at least in these special senses are lacking. And while Nicholas Holland and Jr-Kai Yu found hints of some topography of the touch-sensory axons in the larval spinal cord, the arrangement of these neurons differed significantly from the head-to-tail topographic mapping that is present in the central nervous system of vertebrates.[28] And most sensory pathways to the brain of larval amphioxus have only one or two neurons leading to just one processing center,[29] so this would seem to be too few when compared to ES3 sentient animals.

Finally, despite the clear advancements in their escape behaviors described above, when we look at the neurobiological infrastructure of its brain, while there are glimmers of a midbrain, the dien-mesencephalon, and even some telencephalon,[30] in the next chapter, we will see how far short of the vertebrate condition their brains are when compared to ES3 animals in which sentience is clear.

So while there is no question that amphioxus—especially when compared to ES1 organisms—represents a significant advance on the pathway to sentience, it also falls far short of the ES3 vertebrates and other animals that I discuss in the next chapter. On this basis, I conclude that amphioxus appears to be, in most respects, somewhere between ES1 and ES3 organisms, what I would consider to represent more neurobiologically evolved presentient brains and behaviors.

Sea Slugs

The argument against sea slug sentience is that they have relatively few neurons, few senses, simple neural circuits, relatively small cerebral ganglia, short sensory hierarchies, and they lack a fully developed neural affective infrastructure. Also while their nervous systems do demonstrate specialized neurons and some basic neurohierarchical features, these are a long way away from the brains of ES3 coleoids whose incentive circuits and premotor circuits have become much more neurobiologically complex with innumerable specialized neurohierarchical levels and subcenters. And while their sensing and escape behaviors represent an advance over ES1 organisms, they still show relatively simple approach-avoidance mechanisms when compared to sentient ES3 animals.

On the other hand, their approach-avoidance behaviors are sufficiently complicated that they are beyond what is normally considered reflexes and what we see in ES1 organisms. So as Hirayama and Gillette[31] suggest, these approach-avoidance circuits in *Pleurobranchaea* are a potential bridge from (their word) *precognitive* yet relatively complicated multistep approach-avoidance decision making mechanisms to more clearly sentient behaviors:

> Finally, the interactions of the goal-directed feeding network of the mollusk with its turn network form a simplest decision module for approach/avoidance, acting at a precognitive level in this solitary, cannibal predator. The simple module forms a potentially fundamental type of core circuitry around which the more complex neuronal circuit functions of valuation and comparison are elaborated in the social vertebrates. As such, it can provide a useful starting point for considering the evolution of more complex systems, and it invites future modeling for adding neural and behavioral complexity. (Hirayama and Gillette, 2012, p. 121)

Thus, overall, sea slugs are more evolved on the sentient lines than *C. elegans*, but more on a par with and perhaps even beyond amphioxus. But relative to coleoids such as octopus and squid, they are at a presentient (ES2) level.

Conclusions

While ES2 organisms undoubtedly are *sensing,* and their sensing capabilities are an advance over that seen in ES1 microorganisms, they are missing many critical features of ES3 sentient organisms. These missing features include a lack of the capacity to create *sensory mental images,* and the lack of the affective infrastructure required to create *sentient interoceptive-affective feelings.*

However, there are clear advances regarding the degree of nonreflexivity of their behaviors. And while these reflexes do not show sufficient evidence of sentience, they quite clearly represent an essential "royal road" to its creation. For one thing, they display the required speed of processing and connectivity that higher neurohierarchical levels require for the emergence of sentience. Second, they set up nonsentient but increasingly neurobiologically complex and differentiated adaptive exteroceptive responsiveness that come to fruition in the ES3 animals that we look at next.

7 Emergent Stage 3: Animals with More Neurobiologically Evolved Nervous Systems and Brains—the Emergence of Sentience

Let's summarize the three stages so far. To begin with, the pathway to sentience begins with life, and the emergent features of life continue to play a role in the eventual emergence and evolution of sentience.

Single-celled ES1 organisms bring to the table these features of life as well as basic capacities for reflexive sensing and approach-avoidance responding to the external environment. But these microorganisms while sensing are *nonsentient*.

Presentient ES2 animals have increasingly neurobiologically evolved brains that allows for more sophisticated sensing and adaptive behaviors. But I propose that these animals are *presentient* in that they lack the neural infrastructure for sensory mental images, their neurohierarchical integration is limited or absent, their affective infrastructure is poorly specialized, and their approach-avoidance behaviors, while a clear advance over ES1 single-celled organisms, for the most part are reflexive in nature especially when compared to sentient ES3 animals.

Sentient animals at *ES3* possess brains that are sufficiently neurobiologically evolved that these animals progress from *sensing to sentience*. This is a truly monumental transition that marks a major shift in how animals experience their world and themselves.

The criteria for judging the presence of sentience at ES3 are outlined in chapters 3 and 4, and the neurobiological features that make the emergence of sentience possible at ES3 are summarized in table 7.1. At this level, the animal has clearly delineated neurohierarchical pathways that are capable of creating centralized topographical maps from different senses (e.g., vision, touch, hearing) and these specialized pathways can create *exterosensory mental images*; the animal displays the presence of integrated *interoceptive-affective sentience* as indicated by the presence of the appropriate neural

Table 7.1
Emergent stage 3: Animals with more neurobiologically evolved brains and sentience

First appearance 560–520 million years ago.

Animals at this level: all vertebrates, coleoids (octopus), all arthropods including insects and decapods (such as crabs), and onychophorans (velvet worms).

Novel neural structures and processes supporting the emergence of sentience

- Brain with increased number of neurons (approximately > 100,000).
- Many differentiated neuronal subtypes.
- Elaborated sensory organs with image-forming eyes, receptor organs for touch, hearing, smell, etc.
- More fully developed neural infrastructure for affect and pain.
- Expansion of neural hierarchies with extensive neural interactions.

Novel emergent features related to sentience

- Centralized topographical maps create exteroceptive "sensory images."
- Centralized positive and negative valenced (sentient) affects including pain beyond nociception.
- Increasingly nonreflexive (volitional) actions and globally directed behaviors.

(Adapted from Feinberg and Mallatt 2020)

infrastructure and/or behaviors. And, additionally, both of these first two criteria involve behaviors that are *nonreflexive* and hence truly sentient.

To begin the analysis of ES3 animals, let's go back to the *general biological emergent features* as enumerated in tables 2.1 and 2.2 and compare them to the emergent stage 3 features enumerated in table 7.1. We want to see how the criteria for sentience proposed in chapter 3 relate to these principles.

Sentience Is an Emergent Feature That Is an Aggregate System Feature of Interacting Parts

First, it goes without saying that the view presented here is that sentience is built up from an ever-increasing number of biological and neurobiological "parts" that are *individually* nonsentient. But neurohierarchical organization in neurobiologically complex brains allows for the proposed criteria for sentience including exteroceptive mental images and sentient nonreflexive interoceptive-affective feelings. Thus, I propose that ES1 and ES2 organisms do not have the component *parts* nor their *interactions* between such *parts* that would enable the emergence of sentience.

Support for this view comes from what the three-stage model reveals about sensing versus sentience. As basal emergent processes, simple (nonsentient) sensing can occur in ES1 animals with modest hierarchical organization without any neurons, and in presentient ES2 animals with relatively few intervening neurons such as *C. elegans* and amphioxus; or in even more advanced yet still presentient animals such as *Aplysia* and *Pleurobranchaea* that have more neural hierarchies but relatively more neurobiologically simple brains and largely reflexive responding.

But while these nonsentient and more evolved presentient animals can sense and respond adaptively to their environments, their sensing behaviors are comparatively limited and their associated behaviors remain relatively reflexive (table 6.1). The evidence and criteria for sentience only emerges with increases in the *neurobiological complexity* that is made possible by more *neurobiologically evolved brains* (table 7.1).[1]

The advancements in neurohierarchical organization in ES3 animals also allows for a substantial increase in the sheer number of *differentiated neurons* (table 4.1). Sentient brains have the most neuron *types,* and this differentiation of neurons and their interactions exponentially increases the potential for novel emergent features. So, for instance, ES3 sentient animals—vertebrates, arthropods, cephalopods—have more advanced and better differentiated sensing capabilities when compared to the simple photoreceptors, mechanoreceptors, and chemoreceptors of ES1 microorganisms or even ES2 presentient animals. These ES3 advances include features such as image-forming eyes and other sophisticated neurohierarchical subsystems that subserve hearing, taste, smell, and so on, as well as an expanded affective infrastructure. The same progression applies to the explosion in the differentiation and specialization of different brain regions when compared to nonsentient or presentient brains.[2]

Emergent Features Are Processes Created by the Dynamic Interaction of the System's Parts

At this level, we also see more clearly why the view of emergent system features as *processes* is so crucial. While this is a feature of emergent systems in general, this is also especially important for our understanding of the emergence of sentience since both life and sentience are both embodied *processes.* Here is how evolutionary biologist Ernst Mayr[3] said it:

As far as the words "life" and "mind" are concerned, they merely refer to reifications of activities and have no separate existence as entities. "Mind" refers not to
an object but to mental activity and since mental activities occur throughout
much of the animal kingdom (depending on how you define "mental"), one
can say that mind occurs whenever organisms are found that can be shown to
have mental processes. Life, likewise, is simply the reification of the processes of
living. Criteria for living can be stated and adopted, but there is no such thing
as an independent "life" in a living organism. The danger is too great that a separate existence is as assigned to such "life" analogous to that of a soul. . . . The
avoidance of nouns that are nothing but reifications of processes greatly facili-
tates the analysis of the phenomena that are characteristic for biology. (Mayr,
1982, p. 74)

In fact, many years before, William James[4] already realized that what he
referred to as "consciousness" is a *function*, not an *entity*:

To deny plumply that "consciousness" exists seems so absurd on the face of it—
for undeniably "thoughts" do exist—that I fear some readers will follow me no
farther. Let me then immediately explain that I mean only to deny that the word
stands for an entity, but to insist most emphatically that it does stand for a function. There is, I mean, no aboriginal stuff or quality of being contrasted with
that of which material objects are made, out of which our thoughts of them are
made; but there is a function in experience which thoughts perform, and for
the performance of which this quality of being is invoked. That function knows.
"Consciousness" is supposed necessary to explain the fact that things not only
are, but get reported, are known. Whoever blots out the notion of consciousness
from his list of first principles must still provide in some way for that function's
being carried on. (James, 1904, p. 478)

What is also essential here is that sentience, like life, is physiologically
an *embodied process* and that "consciousness," as per James, is part of what a
living brain *does*. All the general features of biology that we have considered,
and all the special features of sentient brains that we have identified are
functional features of particular living embodiments. This will be critical for
our understanding of the personal nature of sentience that I address later on.

Hierarchical Systems Increase Emergent Properties

Many of the aforementioned emergent features rely upon neural hierarchies
(tables 2.2 and 4.1). This helps explain why all sentient animals have neural hierarchies in all sensory domains. For example, the lower levels that
receive sensory input influence the higher brain levels that in turn dictate

motor output, and they do so far more extensively than in the more reflex-dominated nervous systems of nonsentient organisms and presentient animals.

Emergent Properties Are Novel in Comparison to the Properties of the Parts That Create Them and Their Interactions

Finally, probably the best example of the novelty of emergent features at this stage is sentience itself. This is because sentience fits all the criteria for a biologically emergent property (tables 2.1 and 2.2): *it is a novel aggregate system process that is derived from a neurobiologically complex hierarchical system of living nervous elements, with its novelty attained through the addition of a variety of neural features.*

Another way of saying this is that when we compare the three stages of sentience, we find that the later neurobiological features that create sentience are ideally suited to increase emergent novelty. These are the evolution of enormously increased numbers and differentiation of neuronal subtypes and neural hierarchies that have large numbers of tightly and reciprocally connected neural levels that vastly increase the *enhanced aggregate functioning* that is required for emergence to operate (tables 2.1, 2.2, and 4.1).

Four Phylogenetic ES3 Lines to Sentience: Vertebrates, Coleoids, Arthropods, and Onychophorans

The earlier Feinberg–Mallatt model proposed that there are, at a minimum, three phylogenetic lines of animals that fulfill the primary criteria for the emergence of sentience: all vertebrates, coleoids (octopus, squid, and cuttlefish), and all arthropods including insects and decapods such as crabs.[5] And now I propose that another evolutionary line that meets the criteria for sentience can be added: onychophorans (velvet worms). An updated simplified phylogenetic tree of these sentient lines is shown in figure 7.1. Here is a summary of that analysis.

Vertebrates
As discussed earlier, amphioxus is considered the phylogenetically closest living proxy for the vertebrate line. While we find that sensing and behavioral capacities of amphioxus (ES2) have advanced significantly from ES1

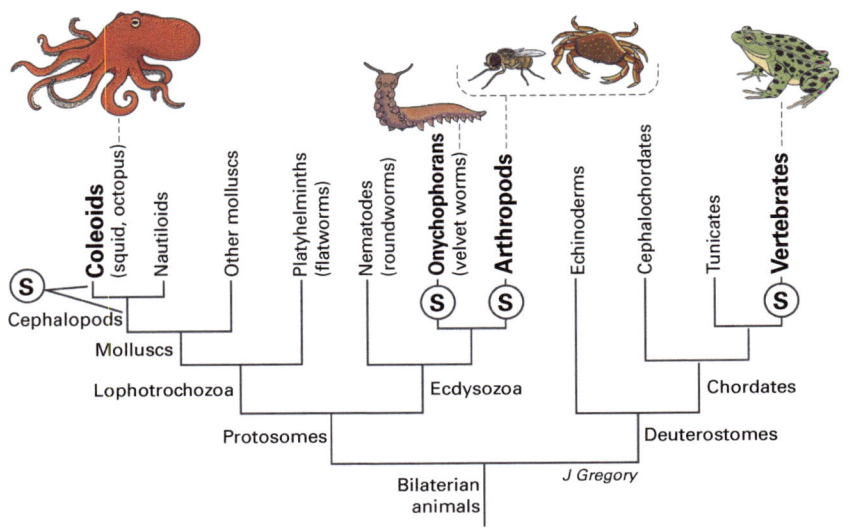

Figure 7.1
A simplified phylogenetic tree of the emergence of sentience in four different lines of sentient animals. On the left, the two leaders extending from the S means that it could not be determined whether sentience evolved in the first cephalopod mollusks or else in the coleoid ancestor of squid, octopus, and cuttlefish. A similar uncertainty applies to the division between onychophorans and arthropods. (Adapted from Mallatt and Feinberg, 2021.)

organisms, they still lack the neurobiological complexity and nonreflexive responding of sentient ES3 animals including the vertebrates.

Exteroceptive sentience: Mapped neural representations and the emergence of sensory mental images   As discussed in chapter 3, to infer the presence of sentient sensory mental images, there needs to be the sensory apparatus and the neural infrastructure in the brain that is capable of creating *mapped neural representations* of a sensory domain. These are the requisite neurobiological features that allow the emergence of sentience from sensing.

It turns out that even the most basal vertebrates possess the neural infrastructure for mapped sensory representations and this is in fact a feature *of all vertebrate brains*. In mammals, the main maps are in the cerebral cortex, and in birds, they are located in the correspondingly enlarged parts of the cerebrum, although they are in somewhat different relative locations in these two groups. However, in more basal vertebrates, for instance fish and amphibians, a midbrain structure called the "optic tectum" contains a finely detailed, point-by-point map of the sensed external environment, mostly from visual inputs but also from the hearing, touch, and balance pathways and it is the region that Mallatt and I proposed is responsible for the mapped representations and mental images for senses such as vision, touch, and hearing. We proposed that the brain site of image-based consciousness shifted from the tectum of more basal vertebrates to the cerebral cortex during the evolution of mammals. However, in vertebrates, the tectum receives no direct input from the smell pathway, so *smell* perception would be performed in the cerebrum in all vertebrates.[6]

There are also some clear neurobiological factors that help explain the dramatic emergence of exterosensory sentience between an ES2 animal such as amphioxus and the basal ES3 vertebrates. This can be traced to the evolution of the vertebrate eye (known as a "camera eye") in the first fish. A camera eye forms a photograph-like image on the retina—*retinotopic image* referred to in chapter 3—and this can be integrated with information from the other senses (for touch, sound vibrations in the water, etc.) that is made possible by the connectivity that is a cardinal feature of ES3 brains.

The evolution of the camera eye is actually quite complicated and amazing. Its development requires a focusing *lens* and the lens of the vertebrate eye develops from embryonic structures that are unique to vertebrates called "ectodermal placodes" that in conjunction with another vertebrate-only embryonic tissue called the "neural crest," develop into all of the special, sentience-associated sensory structures that create mapped neural representations. As proposed in chapter 3, these mapped neural representations are the basis for exteroceptive image-based sentience that distinguish the sentient vertebrates from, for instance, animals that I propose are presentient such as amphioxus.[7] These evolutionary events mark a critical turning point in the creation of exteroceptive sensory consciousness that evolved in early vertebrates over 520 million years ago. The critical nervous system structures of vertebrates and some other ES3 animals are illustrated in figure 7.2.

The neural infrastructure for interoceptive-affective sentience   The affective system of the vertebrate brain—when compared to the tighter somatotopy of exterosensory systems—is more diffuse with innumerable parts that cross communication to promote create integrated emergent functions. But in order for both types of sentience to emerge, they must share the features

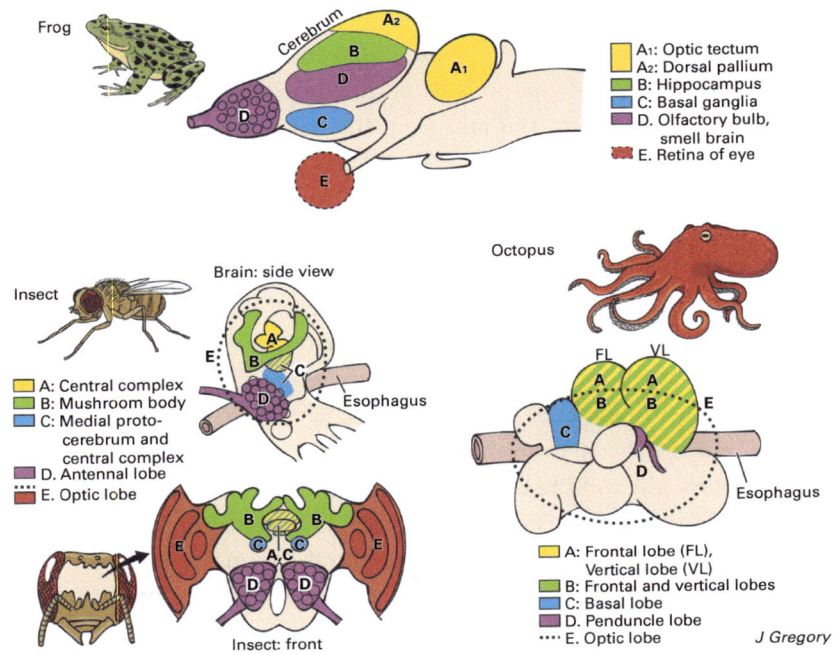

Figure 7.2
A comparison of the brains of three different lines of some proposed sentient animals. Pictured are the brains of a frog (vertebrates), an insect (arthropods), and an octopus (coleoids). Regions with similar functions for sentience are marked similarly in the three kinds of brains. Despite the similarities, the three brains evolved independently of one another. (A) image-based sentience; (B) memory; (C.) premotor center; (D) smell processing; (E) visual processing. (Reprinted from Feinberg and Mallatt, 2018a.)

such as neural hierarchical pathways, reciprocal connections between centers and levels, and neural specialization of parts and centers that are outlined in tables 4.1 and 7.1.

Mallatt and I previously did a comprehensive review of the neuroanatomy that is required for interoceptive-affective sentience including valenced (positive and negative) affects including pain and fear-like responses across all vertebrates. We found that all species including bony fish, amphibians, reptiles, birds, and mammals had nearly solid or complete evidence for the required neural structures.[8] So for vertebrates, the evidence for the requisite neural infrastructure for interoceptive-affective sentience is clear (figures 3.2, 7.2, and 7.3).

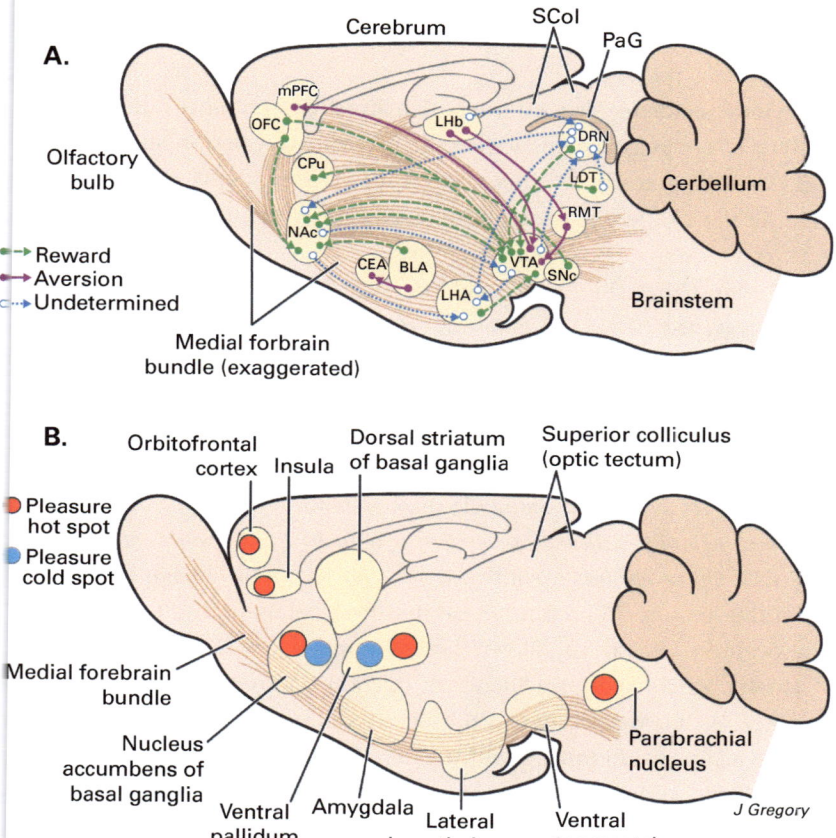

Figure 7.3

Regions of the rodent brain that are associated with interoceptive-affective sentience. Both parts of the figure show the medial forebrain bundle of fibers (light gray), which interconnects the affective regions. (A) shows the major affective centers and their interconnections by valence neurons that code reward or aversion, from mouse studies. (B) shows the pleasure hot spots (red dots) and cold spots (blue dots) in some of the affective regions of the rat brain. These spots were found by applying opioid drugs to these regions and then seeing if this increased or decreased the rat's facial expressions of pleasure when it tasted sugar. Key for part A: BLA: amygdala (basolateral part); CEA: amygdala (central part); CPu: caudate and putamen parts of the basal ganglia; DRN: dorsal raphe nucleus of reticular formation; LDT: laterodorsal tegmental nucleus of reticular formation; LHA: lateral hypothalamus; LHb: lateral habenular nucleus; mPFC: medial prefrontal part of cerebral cortex; NAc: nucleus accumbens of the basal ganglia; OFC: orbitofrontal part of cerebral cortex; PaG: periaqueductal gray; RMT: rostromedial tegmental nucleus of reticular formation; SCol: superior colliculus (optic tectum); SNc: substantia nigra pars compacta of the basal ganglia; VTA: ventral tegmental area.

(Part A is modified from figure 3 in Hu 2016, and part B is from figure 1 in Berridge and Kringelbach 2015. Adapted from Feinberg and Mallatt, 2018a.)

Behavioral evidence for interoceptive-affective sentience In our prior analysis of behavioral evidence for the presence of valenced affect, Mallatt and I used parameters for interoceptive-affective sentience (table 7.2)[9] that were similar to the behavioral criteria for sentient pain proposed by Birch et al. and Crump et al. (table 3.1).

We found that there was solid evidence for *behavioral trade-offs*, such as weighing the benefit of obtaining food versus the risk of higher predation near the food source, in all the vertebrate groups. *Frustration behavior*, such as aggressive behavior after a reward is denied that would be evidence enduring affects, has been demonstrated in fish, birds, and mammals. And two criteria that also appear in the Birch et al. and Crump et al. criterion *self-delivery of analgesics (painkillers) or of rewards*, and *approaching drugs (e.g., amphetamines, ethanol)* or *preferring to be in a place where one previously received drugs or rewards (conditioned place preference* have been found in all vertebrate groups. Therefore, these first four behavioral criteria indicated interoceptive-affective in all vertebrate groups.

Finally, we looked for evidence of operant (instrumental) conditioning (chapter 3) that was similar to the Birch et al. and Crump et al. criterion 7. Here again, we found that all the vertebrate groups—fish, amphibians, reptile, birds, and mammals—possessed operant learning capacities.

Reflexive versus nonreflexive behaviors Note that all the behaviors listed in table 7.2 as well as criteria 4–8 are nonreflexive as defined in chapter 3. Another useful comparison between ES2 and ES3 behaviors as far as distinguishing reflexive sensing versus nonreflexive sentience is concerned is to evaluate the degree of the nonreflexivity of avoidance-escape behaviors in ES2 versus ES3 animals.

Sentient escape behaviors are intrinsically valenced in that they allow the animal to avoid pain or physical harm. Recall in chapter 6 that I reviewed

Table 7.2
Behavioral evidence in vertebrates for interoceptive-affective sentience

1. Behavioral trade-offs, value-based cost/benefit decisions
2. Frustration behavior
3. Self-delivery of pain relievers or rewards
4. Approaches reinforcing drugs/conditioned place preference
5. Operant conditioning with positive or negative outcomes

(Adapted from Feinberg and Mallatt, 2018a)

the work of Lacalli on these behaviors in the cephalochordate amphioxus, an animal that is on the vertebrate phylogenetic line. Although the escape behaviors of amphioxus (figure 6.4B), when compared to ES1 organisms, represent a significant advance on the pathway to sentience, when compared to ES3 vertebrates, it becomes even more clear why these behaviors are most appropriately viewed as presentient.

For instance, let's compare some of the escape behaviors of the vertebrate frog with its ancestor amphioxus. In an ingenious experiment, Bulbert et al. studied the escape response in the ground-dwelling túngara frog *Engystomops pustulosus*. They first propose that it would be advantageous for the frog as prey to vary their escape response depending upon whether the predator was terrestrial—in this case a snake—that was approaching from the ground versus a bat that was attacking from the air.[10]

Presenting the frogs with models of the two predators, the frogs indeed consistently showed different responses depending upon the angle of the attack. So when the frogs were presented with the snake model on the ground, they *fled away*, but in contrast they *moved toward* the bat models. In the latter case, the authors interpreted the frogs' response to the bat model as effectively undercutting the bat's flight path, thus making the attack less effective. They concluded that their results revealed that the frog employed an *adaptive flexibility of strategies* in their different escape response (figure 7.4).

Summary: Emergence and sentience in vertebrates In summary, using the criteria that I presented in chapter 3, the evidence is that all vertebrates,

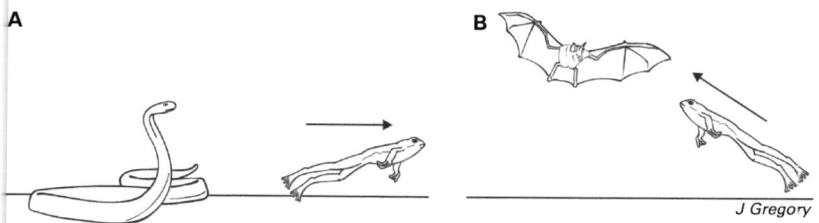

Figure 7.4
The complex escape behavior of the túngara frog. Bulbert et al. (2015) examined the escape responses of the ground-dwelling túngara frog to see whether their escape response varied according to the method of attack: from the ground (terrestrial; snake) versus air (aerial; bat) predators. They found that the frogs *fled away* from the snake models (A) but actually moved *toward* the bat models (B), the latter undercutting the bat's flight path. The authors conclude that there is substantial adaptive flexibility of strategies in the frog's escape response.

even more basal amphibians such as the frog, show all the emergent features of ES3 animals including a full array of senses, an affective neural infrastructure, and nonreflexive sensing and affective behaviors. The presence of these features supports the proposed progression from nonsentient to presentient to sentient organisms.

Coleoid Cephalopods (Octopuses, Squid, Cuttlefish)

The *coleoids* is a group of *cephalopods* that includes the octopuses, squids and cuttlefish. Of these cephalopods, the octopus has received the most attention with reference to sentience, and the clear emerging consensus is that they are sentient.[11] So here I primarily discuss the octopus.

Nervous System of an Octopus

The anatomy of the octopus brain is unique (figure 7.2). By some estimates, its nervous system contains between 170 and 500 million neurons of which about 50 million are in the brain, making it one the largest among invertebrates. In general terms, the nervous system of the octopus is comprised of three divisions: the *brain, the optic lobes,* and the *arm nervous system.* The "brain" includes more than thirty differentiated lobes that are fused together. These are connected to the periphery of the nervous system by nerve trunks that are connected to the animal's arms and other parts of its body.[12]

Exteroceptive sentience: Mapped neural representations and the emergence of sensory mental images   Of all cephalopod mollusks, the octopus has the largest population of sensory receptors. The optic lobe, the largest of the fused central ganglia, contains as many as 65 million neurons. The optic lobes, in addition to processing visual input, plays a significant role in motor control and memory functions.[13]

In both the octopus and squid, strong evidence of retinotopic organization is emerging. For the octopus, Pungor and colleagues recently reported retinotopic visual processing in their optic lobe that they found was similar in some regards to the retinotopicity found in other species.[14] And Chung and colleagues reported clear-cut retinotopic organization in the squid.[15]

So the evidence for the capacity to visual mental images and hence visual exteroceptive sentience in these cephalopod species is solid. However, the evidence for other domains of somatotopy is less clear.[16]

The neural infrastructure for interoceptive-affective sentience Despite strong behavioral evidence for interoceptive-affective sentience in cephalopods and especially the octopus, there is less known as far as direct evidence regarding its neural infrastructure. The presence of nociceptors in the octopus is well documented[17] but currently, the higher brain regions that could support affective sentience need to be inferred.

In cephalopods, the "highest" brain regions that are most often suggested as supporting cognition, memory, and potentially sentience are grouped by various names, but are most commonly referred to as the "frontal-vertical lobe" or more simply the "vertical lobe" (figure 7.2).[18]

Shigeno et al. proposed that the cephalopod frontal-vertical lobe can be compared to the vertebrate fore- and midbrain including the pallium (cortex), hippocampus, and amygdaloid complex; the latter structures that in vertebrates play significant roles in affective functions.

Regarding the homology between these structure in cephalopods and vertebrates, Shigeno et al. [19] offer this interpretation:

> The reason for the deep homology between the vertebrate pallium and the cephalopod vertical lobe system—whether derived from a common ancestral plan or convergently evolved—remains uncertain, but the cephalopod vertical lobe is the best candidate for vertebrate pallium analog within the molluscan lineage (Young, 1991, 1995). (Shigeno et al., 2018)

And there is also emerging direct evidence regarding what brain structures are responsible for sentient pain in the octopus. In a recent paper, Robyn Crook, in order to assess what and how nociceptive activity in the arm could be transmitted centrally the brain, made electrophysiological recordings from the *brachial connectives*, which connect the nerve cords from the arms to the central brain. In brief, the injection of a bolus of pain producing acetic acid in an arm of the octopus resulted in a prolonged period of sustained electrical activity at several sites within the connective. However, after the injection of pain-suppressing lidocaine at the site of the acetic acid injection in the arm, the evoked nociceptive induced activity in the connective was completely abolished, thus providing evidence for the connections between nociceptors and integrative brain regions (Birch et al. criterion 3).[20]

Behavioral evidence for interoceptive-affective sentience There is good experimental evidence that octopuses are capable of associative learning. Birch and colleagues reviewed nonreflexive learning (their criterion 7) in octopods and concluded that there was clear-cut scientific consensus among cephalopod researchers that octopods and cuttlefish have associative learning capabilities.[21]

In the same Crook paper mentioned above, the author evaluated how pain and pain relief influenced the octopus *conditioned place preference*. In this paradigm, the animal is given a choice between environments that are associated with, in this case, pain caused by acetic acid, and then the relief from this pain via the application of the topical anesthetic lidocaine.

First, Crook established which of three chambers a particular octopus preferred in the *absence* of punishment or reward. This was followed by trials where the initially preferred chambers were paired with an acetic acid injection, and as expected, the octopuses spent significantly less time in their initially preferred chamber when compared with octopuses receiving only saline. Then, in trials where octopuses received lidocaine over an area of prior injection (either saline or acetic acid), octopuses preferred the chamber paired with lidocaine only if they had previously been given the acetic acid injection. This sequence of responding to pain and the relief of pain that was linked the animal's place preference provided strong evidence for goal-directed behavior that was motived by pain sentience.

This experiment would support features within Birch et al. criteria 4, 7, and 8. Also, returning to the Birch et al. criteria, their analysis also concluded that there was high or very high behavioral evidence for criteria 4 and 6–8 (table 3.1).

Other nonreflexive affectively motivated behaviors   When we compare the relatively reflexive behaviors of the sea slug *Pleurobranchaea californica* that we discussed in chapter 6 (figures 6.6A and 6.6B) with the neurobiologically complex behaviors of cephalopods, there is a clear and dramatic increase in complicated and nonreflexive behaviors in the latter.

The list of proposed neurobiologically evolved and flexible behaviors in cephalopods is very long and the reader is referred to some of the sources.[22] However, Schnell and colleagues recently provided an excellent and extensive review and of these behaviors that can serve as solid examples of their nonreflexive behaviors. Here, we first briefly highlight some of these that cephalopods show in their natural habitats.

Many of the behaviors that Schnell et al. (2021)[23] discuss in their review are considered examples of the *behavioral flexibility* that would in general make them increasingly nonreflexive in nature: "The suite of cognitive attributes exhibited by cephalopods has likely facilitated their remarkable behavioral flexibility, enabling them innovatively to modify their behavior within various foraging, anti-predatory, and mating contexts" (Schnell et al, 2021, p. 168).

There are many cephalopod behaviors that combine their sensing capabilities and behavioral flexibility. One remarkable example is the cephalopod's sophisticated camouflage capacities that are due in part to the presence of *chromatophore organs* that are located in their skin that allows them to change colors and blend into the environment. Of note, the chromatophores are controlled by a *neurohierarchically* organized set of lobes within the cephalopod brain.[24]

> The chromatophores are controlled by a set of lobes in the brain organized hierarchically. At the highest level, the optic lobes, acting largely on visual information, select specific motor programmes (i.e. body patterns); at the lowest level, motoneurons in the chromatophore lobes execute the programmes, their activity or inactivity producing the patterning seen in the skin. In Octopus vulgaris there are over half a million neurons in the chromatophore lobes, and receptors for all the classical neurotransmitters are present, different transmitters being used to activate (or inhibit) the different colour classes of chromatophore motoneurons. A detailed understanding of the way in which the brain controls body patterning still eludes us: the entire system apparently operates without feedback, visual or proprioceptive. (Messenger, 2001, p. 473)

Cephalopods can also control the papillae of their skin to alter their texture. These combined capacities allow them to change their appearance to both match and blend into all sorts of environmental objects such as rocks or floating algae. This degree of mimicry has enabled some species of octopus to take on the appearance of sponges, a number of fish species, and even sea snakes.[25]

There is yet another different type of behavior that even more clearly shows sentience in an octopus species. Finn and colleagues, over a nine-year period, studied the behaviors of more than twenty Veined Octopus (*Amphioctopus marginatus*) individuals. These octopuses were observed to be occupying empty coconut shell halves, gastropod shells, or even human refuse. When the animals were flushed from the shells by the observer divers, the octopuses reoccupied the shells. Perhaps the most remarkable behavior was that individual octopuses were observed to actually carry the stacked half coconut shells with them for as long as twenty meters for later use. The authors go so far as to interpret this behavior as a form of tool use. However the behavior is characterized, there is no question that this complicated and multistep sequence of actions are *motivated* toward the future goal of obtaining a good shell to inhabit and cannot be reflexive in nature.[26]

Summary: Emergence and sentience in coleoids   In summary, there is solid evidence that cephalopods are sentient ES3 animals with well-developed sensory systems with, at a minimum, somatotopy in its visual system, good but not absolute evidence for the affective neural infrastructure pain sentience, and a range of nonreflexive purposeful behaviors. Collectively, these behaviors support a clear emergent progression from their ES2 ancestors.

Insects

Nervous systems of insects   Among the arthropods (e.g., insects, spiders, crabs, etc.), insects are among the most studied with reference to the presence of sentience. So we will start with them.

The relatively small number of neurons in the insect brain (< 100,000 to 1 million neurons) is considerably fewer than the multimillions in the brains of vertebrates.[27] Yet despite this, the insect brain shows many of the features that are required for the emergence of sentience.

Exteroceptive sentience: Mapped neural representations and the emergence of sensory mental images   The insect brain is quite clearly well differentiated as required for the emergent features of sentient brains that I have proposed (figures 7.2 and 7.5A). They also possess distance senses that include high-resolution vision, olfaction, hearing, taste, as well as various types of touch mechanosenses. Many of these also show *somatotopic organization*[28] that is a critical neuroanatomical feature of exteroceptive sentience (figure 3.1 and table 7.1). One good example of this is that all of the insect senses feature neurohierarchical multilevel serial sensory pathways from the exteroceptive organs to the brain, and some of these reach a fourth or even fifth synaptic level before reaching the forebrain.[29] In one example of this, Seelig and Jayaraman showed that the visual system of the fruit fly retinotopic pathway reaches the *fifth* level of the visual hierarchy into a brain region called "the central complex" (see below) that is involved in higher-order multisensory processing and behavioral decisions.[30]

A fifth-level retinotopic (somatotopic) is worth noting since it is comparable to what we see in mammals who also have at least fifth-order visual processing pathways. And another factor that supports the emergence of sentience in insects is the presence of *reciprocal interactions*, including inter-actions between the processing centers of the different levels of their sen-sory hierarchies.[31]

Another study provides strong evidence of an insect's capacity to form *sensory mental images*. In an ingeniously devised experiment, Solvi and col-leagues tested whether bumblebees were able to learn a cross-modal recogni-tion test between visual and tactile stimuli.[32] They were able to demonstrate that bumblebees who were trained to discriminate object shapes of cubes from spheres via touch (objects explored in the dark) versus in the light (but prevented from touching the objects) could subsequently distinguish these objects via the alternative sense modality. They conclude:

> Whether bumble bees solve the task by storing internal representations of entire object shapes (cube or sphere) or local object features (curved or flat edge) remains unknown. In either case, our experiments show that bumble bees are capable of recognizing objects across modalities, even though the received sensory inputs are temporally and physically distinct. Bumble bees show a kind of information integration that requires a modality-independent internal representation. This suggests that similar to humans and other large-brained animals, insects integrate information from multiple senses into a complete, globally accessible, gestalt per-ception of the world around them. (Solvi et al., 2020, p. 911)

This study remarkably shows that not only are bumblebees able to form sensory mental images in both visual and tactile domains, but they are also capable of integrating these images to guide a complicated nonreflexive behavior.

The neural infrastructure for interoceptive-affective sentience   The most commonly implicated brain regions that are proposed to be involved in the processing of valenced stimuli and possibly pain are the *mushroom bodies* and the *central complex* that includes structures known as the *fan-shaped bodies* (figure 7.5A).

Starting with the mushroom bodies, these complex structures are known to play an important role in the integration of sensory information, learn-ing, and memory.[33] Relevant to affective experience, the mushroom bod-ies play a critical role in ascribing valance to sensory representations. For example, numerous studies have shown a key role of the *Drosophila* (fruit fly) mushroom bodies in assigning positive or negative valence to extero-ceptive sensory stimuli such as odor.[34]

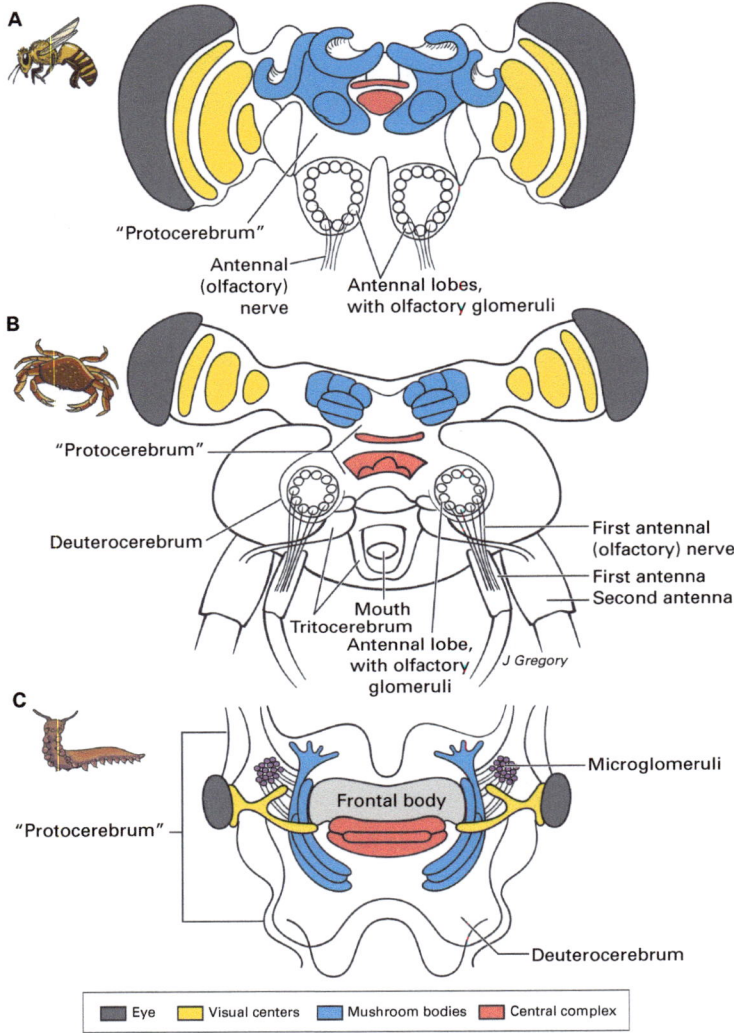

A

"Protocerebrum"

Antennal (olfactory) nerve

Antennal lobes, with olfactory glomeruli

B

"Protocerebrum"

Deuterocerebrum

First antennal (olfactory) nerve

First antenna

Second antenna

Mouth
Tritocerebrum
Antennal lobe, with olfactory glomeruli

J Gregory

C

"Protocerebrum"

Microglomeruli

Frontal body

Deuterocerebrum

| Eye | Visual centers | Mushroom bodies | Central complex |

Figure 7.5

A schematic comparison of the brains of three proposed sentient animals: (A) insects; (B) brachyuran crabs; (C) onychophorans (velvet worms). In insects, the mushroom bodies are known to play an important role in sensory integration and memory. The *central complex* that includes the *fan-shaped bodies* are also implicated in memory functions associated with orientation and path integration. In insects and the proposed sentient crab species (see text), the brain regions that are most commonly proposed to be involved in the processing of valenced stimuli and possibly pain are the *mushroom bodies* and the *central complex*. The functions of comparable regions have not been adequately studied regions in onychophorans but are likely similar. Note that the term "protocerebrum" is the traditional one for what has been recently proposed from developmental genetics as a cerebral volume derived from two distinct neuromeres. (Illustrations are based on for insects and crabs: Strausfeld, 2020, 2021; Strausfeld and Sayre, 2021; Strausfeld et al., 2020; for velvet worms: Martin et al. 2022.)

We saw that the mushroom bodies are important for memory, sensory integration, and exteroceptive sensory valenced responding. Another area of interest for nociceptive processing is a group of midline structures called the "central complex" (figure 7.5A). Like the mushroom bodies, the central complex is also involved in the integration of sensory-motor, memory, and learning functions.[35]

Probably the strongest evidence for the central complex's role in centralized nociceptive or pain processing comes from a study by Hu et al. using fruit flies.[36] In this series of experiments, Hu and colleagues focused on a region of the central complex called the "fan-shaped body (FSB)" that is activated by an electric shock to the insect's leg. Then, after training the flies to avoid the electrified arm of an experimental set-up, they reported that selectively *inhibiting* the activity of a subpopulation of FSB neurons *reduced* the insect's avoidance of both electric shock and noxious heat. And conversely *activating* these neurons via optogenetic stimulation significantly *increased* conditioned avoidance. Also of interest is that these effects were found in both conditioned and innate responding to noxious stimuli.[37] In summary, there is evidence for both the mushroom bodies and the central complex for sentient valenced processing with the greater evidence for pain processing in the central complex.

Finally, Gibbons and colleagues in a comprehensive study of the Birch et al. criteria in insects reported a very high level of confidence that the brains of adult Diptera (flies and mosquitoes) and Blattodea (cockroaches and termites) had *integrated nociception* (see table 3.1, criterion 3), meaning that the animal possessed the requisite neural pathways that connect nociceptor pathways to "integrative brain regions."[38]

Behavioral evidence for interoceptive-affective sentience   There is also solid behavioral evidence for affective sentience such as pain in insects.[39] Two investigations in bees explored whether induced affective states created a "cognitive bias" in bees. A "cognitive biases" is said to occur when an affective state (negative or positive) is found to affect subsequent responding in a matching direction.

In the first study, Batesen and colleagues tested whether honeybees displayed a pessimistic cognitive bias when they are subjected to an anxiety-like state induced by vigorously shaking the bees that was designed to simulate a predatory attack. They found that "agitated" bees indeed were more likely to respond to ambiguous odor stimuli as predicting punishment

and hence avoiding them. They interpreted this finding as the prior negative shaking created a negative cognitive bias and that this represented an
inducement of negative emotional state in the bees.[40]

In a subsequent study, Solvi and colleagues demonstrated the presence
of motivating valence and, hence, affective states, in bumblebees in a well-
known series of experiments that they referred to as the "judgment bias
paradigm."[41] Similar to the Bateson study, in this judgment bias paradigm,
the idea was that by inducing a *positive emotional state* with a sweet reward,
in this case some sucrose solution, the bees would be biased to expect
another positive outcome when confronted with an ambiguous stimulus.

First, the bumblebee subjects were trained to associate different colored
cards with different outcomes—blue (reward) versus green (no reward). Then
in order to see if the bees indeed developed what the investigators considered an *affective* judgment bias, half of the bees received pretest consumption
of sucrose solution that would in theory bias their responding in a positive
fashion with an ambiguous stimulus, this being a blue-green card that was
presented midway between where the initial training cards were positioned.
The actual full experimental paradigm is more complicated with several
additional controls, but the main finding was that the bees that received the
initial pretrial sucrose did indeed develop a judgment bias and took less time
to enter the experimental chamber of the middle ("ambiguous") stimulus.

In addition, in one interesting variation, the investigators further tested
whether the response in the bees was eliminated by the drug fluphenazine
that blocks the neurotransmitter dopamine that is associated with reward processing and emotion in mammals. Indeed, dopamine blockade did eliminate
the bias. In summary, this experiment supports insect affective sentience.

In another recent study also in bumblebees, Gibbons and colleagues
used a *motivational trade-off* paradigm (see table 3.1, criterion 5) that tested
both affective responsiveness to pain, motivation, and nonreflexive behaviors.[42] In a motivational trade-off experiment, an animal is given a choice
(a trade-off) between two competing motivations. In this case, the bees
were initially given a choice between two "high-quality feeders" with a
forty-percent sugar solution and alternative "lower-quality feeders," which
contained lower concentrations of sucrose. The bees of course preferred
the former. The feeders were placed on differently colored heating pads
but during the initial phase of the experiment, no heat (unheated condition) was applied and the bees learned which feeders contained which

solutions. But in a second phase, when two of the high sucrose feeders were heated to noxious temperatures, the bees then *avoided* the noxious feeders when the unheated feeders had high sucrose concentrations, but progressively *increased* feeding from the heated (noxious) feeders when the sucrose concentration at unheated feeders *decreased*. These choices demonstrated a "motivational trade-off" of reward versus nociceptive responses: in other words, *no pain no gain*. The examiners also interpreted this behavior as consistent with the insect's capacity for pain experiences as well as nonreflexive responding. Note also that the motivational trade-off was based on memory, rather than direct sensory experience, which suggests that the bees had some cognitive representation of the competing values of the conditioned stimuli.

Other nonreflexive motivated behaviors   There is a growing literature supporting the variety, complexity, and nonreflexivity of other insect behaviors.[43] With reference as to whether insect behaviors should be deemed sentient, Clint Perry and Lars Chittka provided a particularly valuable analysis on the question of *foresight* and *behavioral flexibility* in arthropod behaviors.[44] In this review, and in some of their other papers, they consider whether the behaviors of many insects is "hard wired." In other words, whether they are *reflexive*:

> The small brains of insects and other invertebrates are often thought to constrain these animals to live entirely 'in the moment'. In this view, each one of their many seemingly hard-wired behavioral routines is triggered by a precisely defined environmental stimulus configuration, but there is no mental appreciation of the possible outcomes of one's actions, and therefore little flexibility. (Perry and Chittka, 2019, p. 171)

Perry and Chittka dispute this claim and provide ample evidence that the behaviors of many insects, particularly those with the largest brains like members of the *Hymenoptera*—a clade that includes their favorite experimental subjects bumblebees as well as ants and wasps—show examples of *complex problem solving, behavioral flexibility, foresight and prediction of the outcomes of one's actions, planning*, and even *tool use* that exceeds what could be considered fixed (reflexive) behavioral routines.[45]

In another remarkable study, Loukola and colleagues did an experiment to investigate whether bumblebees could learn from simply *watching* another bumblebee perform a task.[46] The task for the bumblebees was to move one of three possible balls—the one furthest from the center—into a centrally located specific spot to obtain a sucrose solution reward. They found that the bees that had observed *an actual other bee* perform this task learned the task more efficiently than if they watched what they called a "ghost demonstration" where the ball was moved by a magnet or no demonstration all. Furthermore, not only did the bees show better learning from observing the "live" performance, rather than exactly copying the observed performance, the subject bees solved the task using a more efficient strategy than that observed by moving the ball located closest to the target, even if it was of a different color than the one used in the observed performance. The authors interpreted these learning and cognitive behaviors as good examples of *cognitive flexibility.*

Crabs

Decapod crustaceans: Brachyura and Anomura The decapod crustaceans are a large group of arthropods that includes crabs, lobsters, shrimp, and related species. Within this group there are estimated to be as many as 15,000 species about half of which are crabs. The group known as "true crabs" belong to the infraorder Brachyura. The infraorder Anomura comprise another group of decapod crustaceans that includes hermit crabs and related species. Although they are not considered to be "true crabs," they are a sister group to the Brachyura. We will start with the crabs and after that look at a group closely related to the arthropods called "onychophorans (velvet worms)"; figure 7.6.

Exteroceptive sentience: Mapped neural representations and the emergence of sensory mental images As far as sensory systems are concerned, Brachyura possess complex multisynaptic sensory pathways in multiple

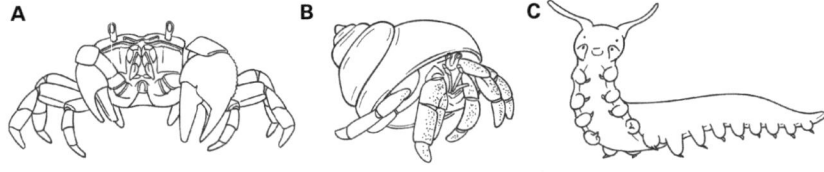

Figure 7.6
Sentient decapod crustaceans and onychophorans: (A) Brachyura ("true crabs"); (B) Anomura (e.g., hermit crabs); (C) Onychophora (velvet worms).

domains including high-grade vision via compound eyes. They also possess multisynaptic and hierarchically organized olfactory, chemosensory, and mechanosensory systems.[47] In a direct comparative study of the Brachyura and Anomura nervous systems, while there are some anatomical differences between the two groups, especially regarding sensory specialization, their nervous systems were found to be largely similar in organization and degree of neurobiological complexity.[48]

The neural infrastructure for interoceptive-affective sentience   A comparative neuroanatomical approach does provide some solid evidence for affective sentience when true crabs are compared to the neural anatomy of insects. For instance, as noted above, the mushroom bodies in insects (figures 7.2 and 7.5A) have been found to play a role in the integration of sensory information, learning, and memory as well as ascribing valence to sensory representations. So the next question is: What are the structures in crustaceans that might play the same role?

Neuroscientist Nicholas Strausfeld, one of the foremost authorities on the arthropod brain, has convincingly argued, along with others, that a brain structure that has come to be known as the "hemiellipsoid body" in crustaceans is actually the homologue of the insect mushroom body.[49] So what some call the hemiellipsoid body, I will take to be the *crustacean* or *decapod mushroom body*.

Figure 7.5 shows the close neuroanatomical correspondence between the insect and the decapod mushroom bodies. So, if we take into consideration *both* the homology of the mushroom bodies between the groups coupled with the behavioral evidence for sentience in insects and true crabs, the evidence for affective sentience in both groups is strong.

Behavioral evidence for interoceptive-affective sentience and nonreflexive motivated behavior   Birch et al. and Crump et al., by their criteria, had high or very high confidence that true crabs (Brachyura) were sentient and high confidence for anomuran crabs.[50] In the case of crabs, in my view, the most compelling evidence for sentience comes from experiments and observations on crabs in naturalistic settings.

Not surprisingly, given their fairly advanced nervous systems, Brachyura species display flexibility and nonreflexivity in their escape behavior that is similar to that shown by the túngara frog that was discussed above.

One of the best studied with reference to their escape behaviors is the Brachyura species *Neohelice granulate*. *Neohelice granulate* (previously *Chasmagnathus granulatus*) is a semiterrestrial burrowing crab that is found primarily along the Southwest Atlantic intertidal zones.[51]

Daniel Tomsic's lab at the University of Buenos Aires has done the most extensive and revealing investigations of the complicated escape behavior of *Neohelice*. As described by Tomsic et al.,[52] the typical first response behavior of the crab when it visually observes the approach of a moving object—referred to as a "looming stimulus"—is to freeze, theoretically to reduce the animal's chances of detection by a potential predator. Then, if the predator advances toward the crab, the animal initiates what Tomsic et al. call a "home run," wherein it runs to the entrance of its burrow, into which, if the threat escalates, the crab enters. If no burrow is available, the sequence is freeze, running away, and finally raising its claws at the threat.

What is of particular interest here is how similar the behavior of this invertebrate parallels the escape behavior of the vertebrate túngara frog:

> Crabs also correct their escape direction according to changes in the observed trajectory of the danger. A crab may move away from a threat in two ways: it can keep the same orientation of its body in space but change the course of locomotion; alternatively, it can rotate and visually fixate the predator with the lateral pole of one of the eyes . . . and then use its preferred sideways style of running to escape from the danger . . . These results demonstrate that while escaping from a visual danger, crabs constantly adjust the speed and direction of the run according to ongoing changes in the flow of visual information. (Tomsic et al., 2017, p. 2319)

The authors also note that while these escape behaviors are largely innate, they are also highly modifiable by numerous variables including by time of day, the current season, and the risk of predation as determined by the relative abundance of predators. Perhaps most significantly, they are also modifiable by a number of learning and memory variables.[53] Thus, the adaptive variability and modifiability of the escape behaviors reinforces the support for their *adaptive flexibility*.

But even more striking evidence for nonreflexive and neurobiologically evolved behaviors can be found in the hermit crab. Robert Elwood, who is Emeritus Professor in the School of Biological Sciences at Queens University Belfast, and colleagues have studied hermit crabs for decades and he offers an argument for sentience in these animals based upon their complex behaviors with gastropod shells that most hermit crabs inhabit.

Hermit crabs have a soft exoskeleton that makes them vulnerable to injury from predators or other crabs. They have adapted to this by inhabiting empty gastropod shells for protection. The diversity of the behaviors in regard to these shells is quite complicated and the reader is referred to Elwood's comprehensive summary,[54] but these include seeking and selecting larger shells as the animal grows, judging shells on the basis of physical fit and weight, and engaging in complicated and multivariable "contests" with other crabs if a housed crab deems the opponent's shell preferable to its current shell.

Elwood also observed a number of behavioral features of hermit crabs that indicate their sentience along the lines of the Birch et al. and Crump et al. criteria, including motivational trade-offs between avoidance of noxious stimuli and other motivational requirements, attending to sites affected by injury, and negative motivational changes toward shells if the crab received shocks there.[55]

But my favorite is what is referred to as a hermit crab "vacancy chain." The general term "vacancy chain" has been widely applied across disciplines, but the fundamental idea is that it occurs when a new "resource unit" is acquired by the lead individual in a chain who then leaves their prior unit unoccupied. Then, a second individual in the chain occupies that unoccupied unit left empty by the first individual thereby in turn leaving their old unit unoccupied and thus making that unit available to the next individual in the chain and so on. In humans, vacancy chains may be created for a range of resources such as vacancies in more desirable jobs, automobiles, consumer goods, and so on, but a common one is housing units.

The term "vacancy chain" was later rather ingeniously applied to hermit crab behavior as they "line up" to get their new gastropod shell acquisitions.[56] Crabs need new shells as they outgrow their old ones or the latter deteriorate over time. Since new shells may not be readily available, there may be a competition for these among the crabs.

Basically, the vacancy chain begins when an individual crab "inspects" a vacant shell to decide if it's a good fit and a desirable replacement for their current shell. If they don't like the shell after inspecting it or "trying it out" for fit, the crab keeps its original shell and a line forms behind it. In one scenario, the crabs form a line in size order largest to smallest in sort of a hermit crab "conga line," as described by Rotjan et al.:

Previous descriptions of hermit crab vacancy chains have noted "queuing" behavior . . . which we define as the formation of one or more size-ordered, linear arrays of hermit crabs in which the largest crab in each line is grasping an empty shell, and each successively smaller crab grasps from behind the shell of the preceding crab. We define a similar behavior, "piggybacking," whereby 2 or more crabs line up (not necessarily in order by size) by grasping the shell of another crab from behind. Piggybacking does not involve a vacant shell, and the lead crab often continues walking with the attached crabs trailing behind it. (Rotjan et al., 2010, p. 640)

I think one reasonable interpretation of individual crab behaviors in vacancy chains is that they represent sophisticated and goal-directed, motivated, and nonreflexive behaviors that are affectively related in the sense that they have a positive *valenced* goal, that being the acquisition of a more desirable shell.

In conclusion, based upon the criteria for sentience that were outlined in chapter 3, I believe that the weight of the evidence supports the view that both Brachyura (true crabs) and Anomura (such as hermit crabs) meet the criteria for sentience because they display a sufficient degree of exteroceptive and interoceptive-affective sentient feelings, and their behaviors including social behaviors are nonreflexive.

Onychophorans (Velvet Worms)

Velvet worms are members of the phylum Onychophora and by most estimations are the closest living relatives of the arthropods from which they diverged as far back as the Cambrian period over 540 million years ago (figure 7.1). Their ancestors were worms that roughly resembled onychophorans. The proposed taxon *Panarthropoda* that includes Onychophora (velvet worms) Arthropoda (e.g., insects, crabs, etc.), and Tardigrada ("water bears") captures the common phylogenetic origins of these three groups.[57]

Nervous system of velvet worms   The nervous systems of velvet worms have at a minimum a bipartite (two-segmented) brain and Martin and colleagues report that the mushroom bodies and central bodies in onychophorans are homologous with these structures in arthropods[58] (figure 7.5C).

Exteroceptive sentience: Mapped neural representations and the emergence of sensory mental images   The sensory systems of velvet worms are diverse. The *mechanosensory* system is based upon protrusions called "papillae" that cover the animal's body and feet. Theses papillae are connected to a "bristle" at the tip that in turn is connected to sensory nerve cells that are responsive to touch. The chemosensory system consists in sensory cells known as "sensills" that are located primarily around the animal's "lips" and its two antennae. Antennal nerve cords are connected to antennal neuropils that project to bilateral olfactory lobes. The latter are complex structures that consist of eighty subunits called "olfactory glomeruli" as well as large macroglomeruli[59] (figure 7.5C).

The *visual system* is composed of simple but nonetheless image-forming eyes (ocelli) that are located behind each antenna. Each eye features a lens, cornea, and retina that is connected to an optic nerve that projects to the animal's brain. However, the eyes of velvet worms are small and have very low-resolution, so their vision is much less acute when compared to the compound eyes of arthropods or the large eyes of vertebrates and octopuses.[60]

The neural infrastructure for interoceptive-affective sentience   The interoceptive and affective behaviors of onychophorans have not been specifically investigated. However, the homologies between the brains of insects, decapod crustaceans, and velvet worms (figure 7.5) makes it highly likely that if studied, then affective behaviors would be found. And these homologies between brain structures coupled with the high degree of complicated nonreflexive social behaviors and interindividual interactions of velvet worms strongly supports their sentience. I consider the latter next.

Behavioral evidence for interoceptive-affective sentience   While the specifically affective behaviors of velvet worms have not been extensively studied, there is ample evidence for neurobiologically evolved and nonreflexive *behavior* in some velvet worm species that is comparable to some other ES3 levels animals. These behaviors are affectively related, nonreflexive, motivated, and goal directed.

For example, in a remarkable observational study of the social behavior of *Euperipatoides rowelli*, Reinhard and Rowell report that the individual animals cluster into groups that consist of as many as fifteen female, male, and young worms.[61] They found that in fact these social groups are close knit and they display a female dominant social hierarchy. They hunt collectively but the dominant female of the group eats alone and before the other members of the group. How the hierarchy is established is of interest:

> Hierarchy within a group is established by aggressive-dominant and passive-subordinate behaviours, the latter leading to tolerance of body contact and aggregation. *Euperipatoides rowelli* from foreign groups, i.e. from different logs, are met

with intense aggression, and individuals rarely aggregate. The reasons for this aggression are not clear, but we suggest that its origins lie in kin recognition. (Reinhard and Rowell, 2005, p. 1)

We detected a strict female dominated feeding hierarchy: the first individual to feed was always a female, and she remained the sole feeder for 45–60 min. This female fed in bouts of up to 10 min, interrupted by short intervals during which she moved around the Petri dish. During her solitary feeding time, males and young of the group remained distant often in physical contact with one another. The other females assumed a 'waiting' position close to the prey, circled it, and occasionally tried to feed. If noticed by the first female, they were attacked by her, bitten, kicked and pushed away. After 45–60 min, the first female tolerated other individuals at the food. Now, the remaining females started feeding in groups, soon followed by the males and young. (Reinhard and Rowell, 2005, p. 3)

In summary, velvet worms have primitive but image-forming eyes and based upon the homologies between their brain anatomy as compared to insects and crabs, they most likely possess sufficient affective infrastructure for sentience. Finally, their behaviors, especially their social behaviors, are so neurobiologically complex and evolved and nonreflexive that I propose that, in fact, velvet worms are sentient.

Summary: Emergent Stage 3 and Sentience

To summarize, there is striking correspondence between the increase in the general features of emergence that occur in *all* neurobiologically complex biological systems and the increase in emergent features that occur in sentient brains at ES3. Indeed, this proposition encapsulates the essence of the theory of Neurobiological Emergentism (NBE): sentience that—despite its unique qualities—has all the properties of an emergent process that are listed in table 2.1; it is a novel *process* that comes from a *neurobiologically complex, neurohierarchical system* of living neural elements, with its emergent novelty attained through addition of system features that are *not present in the system's parts*. The increase in these features is how ES3 animals evolved sentience.

But bear in mind that while the evolution and creation of sentience requires the successive *addition* of novel structures and processes that result in unique emergent features, there is also the *retention* of critical basal emergent features—for instance life and process—that are present and contributory at both lower and higher evolutionary levels. The implications of this for an understanding of sentience is that describing any single biological, neurobiological, or evolutionary factor alone cannot explain sentience. Rather, many variables must be considered and factored into a comprehensive explanation.

There is another interesting feature regarding the evolutionary timeline of the emergence of sentience (figure 4.1). There were approximately 3 billion years of evolution between the emergence of *nonsentient* ES1 single-celled organisms and the first *presentient* ES2 animals; but between presentient ES2 animals and *sentient* ES3 animals, there is only a 10–50-million-year gap! This suggests that between ES2 (Precambrian) and ES3 (Cambrian) levels of sentience, in fact, the pace of the evolution of presentient neurobiological features was dramatically accelerating. One inference from this is that once the emergent neural factors that led to sentience had begun at ES2, the evolution of these novel neural features increased the pace of emergence of sentience in concert with the advances in neurohierarchically advanced brains.

Finally, according to the theory, there is nothing unusual or "mysterious" about the basic physical properties of either the parts of the brains (neurons) that create sentience nor is there anything unexplainable about how the emergence of sentience is consistent with the standard principles of all biological emergence.

But in order to explain how sentience can be a natural, personal (subjective) system feature of certain brains, we need to further elucidate the different ways that have been proposed to explain the notion of "emergent levels." What is actually going on at the micro-macro neurobiological level of brains is exceedingly important for NBE. In the next chapter, we take a look at that critical problem.

One of the major obstacles to unifying the biology of the brain, the levels ES1-ES3 that I have proposed, and the special problems that are posed by the subjective features of sentience is understanding exactly how these three issues are related. This question is especially important for clarifying some philosophical issues about sentience that are discussed in these final sections of this book. The ultimate objective is to provide a scientific scenario that can proceed seamlessly from biology to sentience. And that explanation of sentience must include its personal nature.

This chapter addresses how and where in the brain—spatially and temporally—the emergence of sentience actually occurs. This is one of the most challenging aspects of the problem of the emergence of consciousness in general and the one that has perhaps the most profound implications for our understanding of how "feelings" are created.

Emergence and Causation: Hypotheses and Controversies

Roger Sperry: Levels of Organization and the Problems with "Emergence at the Top"

One important problem that we can identify is the tendency to attribute the emergence of sentience *as occurring at* the "highest" level of brain organization. Since it is generally supposed—correctly in my view—that consciousness or sentience requires a neurobiologically complex hierarchical nervous system, it could intuitively appear that as additional neural levels are added to the neuroaxis consciousness might emerge "at the top" or at the "highest level" of the brain. I call these "emergence at the top" theories of consciousness.

There is a cluster of related philosophical debates about the nature of consciousness that deal with this general line of thinking. In one proto-typical and well-known version, Nobel laureate Roger Sperry[1] proposed an *emergent interactionist approach* in which he argued that the *subjective proper-ties of consciousness* emerge "at the highest levels in the hierarchy of brain organization" and then in a "position of top command" control the mate-rial brain (figure 8.1):

> Consciousness was conceived to be a dynamic emergent of brain activity, neither identical with, nor reducible to, the neural events of which it is mainly composed.

Figure 8.1
Roger Sperry proposed an emergence of consciousness "at the top" of the brain. Sperry's theory proposes that the emergence of consciousness occurs at the "highest" level of neural organization. According to this line of reasoning, it could intuitively appear that as additional neural levels are added to "higher" (more rostral) levels of the neuroaxis, the emergence of consciousness might emerge "at the top" or at the "highest level" of the nervous system or the brain.

Further, consciousness was not conceived as an epiphenomenon, inner aspect, or other passive correlate of brain processing, but rather to be an active integral part of the cerebral process itself, exerting potent causal effects in the interplay of cerebral operations. In a position of top command at the highest levels in the hierarchy of brain organization, the subjective properties were seen to exert control over the biophysical and chemical activities at subordinate levels. It was described initially as a brain model that puts "conscious mind back into the brain of objective science in a position of top command . . . a brain model in which conscious, mental, psychic forces are recognized to be the crowning achievement . . . of evolution." (Sperry, 1990, p. 382)

I think at least one of the reasons that Sperry—and maybe others—were attracted to "emergence at the top" models is because the strict nonnested neurohierarchical properties of exteroceptive feelings (especially visual and tactile pathways) give the appearance that a complete "neural somatotopic model" culminates or emerges *at the top* of these hierarchical sensory pathways in the cerebral cortex; and then it would be logical that the emergence of sentience in most vertebrates and especially primates, should occur at these *highest levels of the brain* (figure 3.1; 8.2 below).

However, not only is this "emergence at the top" an illusion for the exteroceptive pathways that Sperry focuses on, even if it were a valid view for those systems, it would apply much less or in some cases not at all to those aspects of interoceptive-affective processing in which nonsomatotopic feelings (such as emotion) are especially important but not strictly hierarchical in the same way as are exteroceptive pathways. So it should be clear by now why NBE proposes to view sentience as an *emergent multilevel aggregate system property* of neurobiologically complex nervous systems.

So while the biology and evolution of sentience requires the successive *addition* of novel structures, processes, and "levels" that result in novel emergent features, there is also the *retention* of critical emergent factors—for instance life and process—that are present and active at both lower and higher emergent levels. And just as life itself is a basal nonsentient emergent biological feature of all organisms, yet it still plays a critical a role in the eventual emergence of sentience, so do all the astounding number of "lower" and "higher" parts and levels of the nervous system that contribute to the emergence sentience. However, these levels do not function as if they are "layered" one upon another.

Further support for this view comes from the way that basic *sensing* that is present at all emergent stages and as a precursor of sentience progresses to *sentience.* How relatively more basal brains, such as those at ES1, can create surprising varieties of sensing, but for the reasons that I have enumerated, the emergence of sensing does not entail the same "degree" of aggregate system emergent properties that requires the features of sentient brains that I enumerate in table 4.1. Thus, sensing is a beginning, but not the end.

Finally, regarding Sperry's proposal, although he denied that his theory is dualistic,[2] the sort of "mind-brain" interaction that he proposes certainly seems "dualistic" and appears to inevitably lead to some variety of mind-body (psychophysical) dualism. As Sperry says in the above quote, "the subjective properties were seen to exert control over the biophysical and chemical activities at subordinate levels." This is certainly one way of expressing a dualist theory along the lines of the Cartesian notion of the res extensa (the "physical") and res cogitans (the "mental") wherein the subjective mental properties can control or "interact" with the physical brain.[3] We will return to this idea in chapter 10.

"Strong" versus "Weak" Emergence and Downward Causation

Therefore, not only is the "emergence at the top" view *scientifically* wrong, it is also *philosophically* problematic for several reasons. But in my opinion, the most problematic is that this view makes sentience appear as an *immaterial* emergent feature of the physical brain and thus creates a perplexing transition from the *physical* brain to *nonphysical* sentience and vice versa. In other words, this view supports a theory of *dualism* between brain and mind; it does not help explain the explanatory gap, it just makes it more scientifically mysterious.

Recall from the Preface that the difficulty of explaining how sentience could naturally emerge from the brain, and how something that is an "objective" brain process, could be personally (ontologically) "subjective" has led some to the claim that no standard version of emergence ("weak" emergence) could explain sentience. This more radical position is called "strong" emergence.

A major difficulty with strong emergence views of sentience or consciousness—in common with Sperry's view—is whether they can avoid a variety of mind-brain dualism in which consciousness is an "immaterial" feature of brain functions that somehow "pops out" of the nervous system. This contributes to the problems posed by the explanatory gap that we introduced in chapter 1.

Philosopher Antti Revonsuo nicely summarizes a version of this position that resonates with "emergence at the top" theories as well:

> Supporters of strong emergent materialism point to the fundamental differences between the subjective psychological reality and the objective physical (or neural) reality. The former includes qualitative experiences that feel like something and exist only from the first-person point of view; the latter consists of physical entities and causal mechanisms that involve nothing subjective or qualitative about them and exist from the third-person point of view or objectively. Nothing we can think about or imagine could make an objective physical process turn into or "secrete" subjective, qualitative "feels." It is like trying to squeeze wine out of pure water: it is just not there, and there can be no natural mechanism (short of magic) that could ever turn the former into the latter. (Revonsuo, 2010, p. 30)

Note that Revonsuo brings up three of the most important and "mysterious" aspects of sentience. First, there is the question of how sentience emerges. Second, there is the question of how the "material" brain creates something that has an ontologically subjective existence. Third, how can "physical entities'" and "causal mechanisms" that apparently have nothing subjective about them create something that has subjective qualitative "feels."

A version of a strong emergence theory has been proposed by David Chalmers. Interestingly, it confronts similar problems to those faced by Sperry. Chalmers view is in part an effort to explain the "hard problem" of consciousness as discussed in chapter 1. His solution is that there must be something physically "fundamental" about sentience or consciousness.

As he explains it here, he proposes that there must be some fundamental novel physics at work in the creation of consciousness and sentience:

> Strong emergence has much more radical consequences than weak emergence. If there are phenomena that are strongly emergent with respect to the domain of physics, then our conception of nature needs to be expanded to accommodate them. That is, if there are phenomena whose existence is not deducible from the facts about the exact distribution of particles and fields throughout space and tie (along with the laws of physics), then this suggests that new fundamental laws of nature are needed to explain these phenomena. (Chalmers, 2006, p. 245)
>
> We have seen that strong emergence, if it exists, has radical consequences. The question that immediately arises, then, is: Are there strongly emergent phenomena? My own view is that the answer to this question is yes. I think there is exactly one clear case of a strongly emergent phenomenon, and that is the phenomenon of consciousness. We can say that a system is conscious when there is something it is like to be that system: that is, when there is something it feels like

from the system's own perspective. It is a key fact about nature that it contains conscious systems; I am one such. And there is reason to believe that the facts about consciousness are not deducible from any number of physical facts. (Chalmers, 2006, p. 246)

While Chalmers does not describe what these "new fundamental laws of nature" might be, nor does he specify at what level of physics these fundamental forces exist, in the second quote above, in his view, the only case of strong emergence that exists is the "phenomenon of consciousness." In any event, there is little doubt that his strongly emergent hypothesis could lead him to a similar dualist dilemma that was faced by Sperry.

And indeed, Chalmers arrives at a philosophical position he called "naturalistic dualism."[4] In this theory, he proposes that sentient states "naturally supervene" (depend upon) physical systems (such as brains) but at the same time he asserts that mental states are *ontologically distinct* and not "reducible" to physical systems.

In my way of thinking, this is a bit like trying to have your cake and eat it too. Like Sperry, the claim here is that somehow the brain creates emergent mental states yet in some other sense their subjective aspects are independent of those neural states. So his view could be interpreted as arguing that even though mental states are created by the brain, they are in some critical and ontological sense *separate* from the brain.

For instance, let's consider the claim that a "nonphysical" emergent consciousness can have causal effects on the "physical" brain. Just like Sperry's model of the emergence of consciousness "at the top" is an example of a strong "upward" emergence, his *emergent interactionism* or *emergent or downward causation* are early versions of *strong downward causation* (see Chalmers below), the "strong" part being that once consciousness emerges *from* the brain, it then acquires control *over* the brain:

> My perspective in what follows is centered around the contention that this turn-about in the casual status of mental entities requires a shift to a new form of causality, a shift specifically from conventional microdeterminism to a new macromental determinism involving "top-down" emergent control (and referred to variously as emergent interaction, emergent or downward causation, and also as macro, emergent, or holistic determinism—among other labels). If I am correct, emergent determinism is the key to the consciousness revolution. (Sperry, 1991, p. 227)

So, as Sperry put it in the quote cited earlier: "In a position of top command at the highest levels in the hierarchy of brain organization, the subjective properties were seen to exert control over the biophysical and chemical activities at subordinate levels."

This part of Sperry's theory is also a version of a solution to the philosophical problem of *mental causation*. Here is how Revonsuo[5] defines mental causation: "The idea that the mind or mental phenomena have causal powers to change some purely material (e.g., biological or neural) process in the brain" (Revonsuo, 2010, p. 298).

In Sperry's case, it is also of interest how he conceived of downward (mental) causation in terms of micro- and macrodeterminism (e.g., along the lines of the "levels of emergence" that we have been analyzing). Again, the problem with his view is that, ultimately, consciousness mysteriously appears as an *emergent downwardly causal factor* "from the top" of some sort of emergent neural hierarchy.

Along with Chalmers' suggestion that I discussed above, that consciousness may be a "strongly" emergent phenomenon, he also at least considers the possibility that consciousness may have "strong" downward causation effects upon lower-level brains processes:[6]

> To be clear, one should distinguish strong downward causation from weak downward causation. With strong downward causation, the causal impact of a high-level phenomenon on low-level processes is not deducible even in principle from initial conditions and low-level laws. With weak downward causation, the causal impact of the high-level phenomenon is deducible in principle, but is nevertheless unexpected. As with strong and weak emergence, both strong and weak downward causation are interesting in their own right. But strong downward causation would have more radical consequences. . . . I do not think there is anything incoherent about the idea of strong downward causation. I do not know whether there are any examples of it in the actual world, however. (Chalmers, 2006, p. 249)

However, from the NBE point of view, there is no need to posit the existence of "radical" or "strong" downward causation in conscious processes, any more than there is the need to propose the existence of the "strong" upward emergence of consciousness.

My reasoning is that emergence is a natural factor in the creation of sentience just as emergence is a universal feature of all complex biological systems, the main difference between these two being the extent to which neurobiological hierarchies greatly maximize the degree of these emergent processes. But what appears to be an "interaction" between subjective experience and the objective brain is an illusion.

From the standpoint of NBE, the essential point is that in fact the brain and its subjective features are aspects of the *same emergent aggregate system*. In other words, they are *not dissociable*. And sentience, experience, consciousness, feeling, and so on, as personal emergent system features of the brain only have causal powers *within that system*. In other words, the causal properties of sentience with regard to an animal's brain and nervous system are *personal to the individual animal*.

Thus, there can be no "interaction" or "dualism" between the brain and its emergent sentient features. While they are different "aspects" of this system, they are *both aspects of the same emergent system*. In other words, there is no "gap" and no basis for some "causal interaction" between the brain and the "mind" as if they were separate entities.

NBE makes this dualist position unnecessary. First, as far as the NBE model of the biology, neurobiology, and the evolution of sentience is concerned, we can explain the mechanisms of the emergence of sentience via standard biological and evolutionary principles that could apply to *any* emergent process. So there is no need for positing a unique *type* of emergence to account for sentience.

Second, another thing that distinguishes NBE from other emergentist theories is that when these basic emergent principles are applied across nonsentient, presentient, and sentient stages, there is a clear progression in biological and neurobiological stages *that are specifically linked to the emergence of novel features*. In this way, we can trace the emergent progression from basic stage 1 *nonsentient sensing* mechanisms to stage 2 *presentience* that is characterized by significant advances in sensing functions but not quite at the emergent level of sentient animals. And finally, with the advent of advanced brains at stage 3 *sentience*, these animals in general possess all the criteria of sentience including exteroceptive sensory mental images, affective awareness, and nonreflexive behaviors. Importantly, in evolution, this progression is step-wise but neurobiologically and emergently seamless and sentience doesn't suddenly "pop out" of the brain.

These considerations lead me to believe—in contrast to Chalmers—that no new *physical principles* are required to explain the emergence of sentience, and that in actuality nervous systems are *uniquely* capable of maximizing the potential and novelty of emergence. Rather, the emergence of consciousness is simply a matter of the *degree* of standard emergence, and the *unique* properties of neurobiological emergence, but not a different *kind* of emergence. But such a massive *quantitative* increase in unique emergent processes gives the false impression of a *qualitative* explanatory gap between the brain and sentience when there actually is none. I will address in chapters 10 and 11 another of Chalmers' concerns (that I introduced in chapter 1), namely the problem of how to explain the "something it is like to be" from the animal's *personal perspective,* but next I look further into how neurohierarchical "levels" actually relate to each other.

Emergence, the Brain, and Models of Hierarchical Organization

So what alternative models are there that might be more scientifically plausible ideas of the relationship between *neurohierarchical levels of organization* and emergence of sentience?

It turns out that the analysis of levels of organization in general and as it relates to the neurobiology of sentience is a complicated and controversial matter. For instance, in a recent and excellent book on this subject entitled *Levels of Organization in the Biological Sciences,* the editors of this book Daniel Brooks, James DiFresco, and William Wimsatt write that the various theories on the subject of hierarchical "levels" often differ, in some cases dramatically so, along methodological, epistemological, and ontological lines. In their editorial introduction to this book, they explain that while the related concepts of "levels" is often used interchangeably with that of "hierarchies," how these two are related is both complicated as well as specific to the problem at hand.[7] Indeed, Brooks opines that meanings of the concept of "levels of organization" across disciplines and subject matter are so diverse that it is unlikely that there will emerge any universal usage of the term:

> Each experimental system will comprise its own descriptions, guided by distinct investigative questions and interests, even concerning general structures like molecules, cells, and tissues. Instead, 'levels' in scientific research often expresses knowledge heavily contextualized within a particular system of interest, with its attendant methods, techniques, theories, and scientific questions guiding the expressed content of the concept. This in turn means that 'levels' is highly sensitive to the circumstances of its usage. (Brooks, 2017, p. 41)

A full discussion of these more recent models of the levels of hierarchical emergence and all its ramifications in different nonbiological and biological systems is beyond the scope of the present analysis, and the reader is referred to the Brooks et al. volume as well as a number of valuable analyses of the subject.[8] But Brooks et al. and Brooks in an earlier paper raise the issue of a "layer cake" approach to emergence. This issue is of particular importance for our analysis of sentience because it strongly resonates with many of the philosophical issues surrounding the relationship between the brain and consciousness.

As Brooks et al. point out, the older strictly "layered" image of emergent levels, in which each hierarchically higher level is neatly "layered" upon a lower level, has been supplanted by more nuanced approaches. Here is how Brooks in a prior paper[9] described this earlier and less sophisticated *vertically stratified* model of emergence:

> A common idea found throughout philosophy and science, especially the biological sciences, is that the world, or some part thereof, is hierarchically structured into various levels of organization. The image of the world that this idea evokes posits a vertically stratified structure in which the constituents of nature, or even the sciences themselves, are connected together into a graduated continuity. The things found at one horizontal slice of reality somehow "make up" or "are continuous with" the things found at the next slice, and so on. From this basic characterization, various ontological, methodological, or epistemic claims concerning features or patterns in nature and science are said to follow (Wimsatt 1994/2007, p. 214; Potochnik and McGill 2012, pp. 125–126). The exact expression of these basic features, the claims they support, and even the debate(s) to which they belong, vary among sources. (Brooks, 2017, p. 142)

Jaegwon Kim, one of the best know philosophers on the subject of hierarchy and consciousness endorsed a version of the "layered" approach:

> Although the fundamental entities of this world and their properties are material, when material processes reach a certain level of complexity, genuinely novel and unpredictable properties emerge, and . . . this process of emergence is cumulative, generating a hierarchy of increasingly more complex novel properties. Thus, emergentism presents the world not only as an evolutionary process but also as a *layered structure*—a hierarchically organized system of levels of properties, each level emergent from and dependent on the one below. (Kim, 1992, pp. 121–122)

In some respects, Kim's conception of the "material processes" reaching a level of complexity and the emergence of novelty makes perfect sense. But the problem again is the idea of a "layered structure" with reference to nervous systems that create sentience is problematic in many respects that I now analyze.

There Are Myriad Levels and Different Types of Interactions That Can Contribute to the System Emergence of Sentience

If sentience is an emergent system property that is the result of the operation of numerous diverse hierarchical levels of organization—however they are drawn—and these levels are mutually nondissociable, then how are we to describe or determine their relationship to sentience?

Indeed, in my view, much of the debate regarding the relationship between the brain and the emergence of sentience ultimately revolves around this question. But I think that we are now in a position to untangle these issues in a neurobiologically and philosophically coherent way.

According to the neurobiological emergentist model that I propose, sentience ultimately emerges in progressive and successive stages of increasing neurohierarchical complexity and neurobiological novelty at multiple neural levels; and this is achieved as a result of an explosive increase in emergent processes as outlined by the factors in table 4.1. Again note, however, that these are system features and do not emerge *at* a given neural level.

Indeed, it is often overlooked how diverse these hierarchies become. In fact, neural hierarchies come in a vast array of architectures that create emergent properties at multiple scales and feature varied functional properties. This proposal is also of importance for our understanding of an evolutionary model of the emergence of consciousness since these various neurobiological hierarchies are seen in abundance across all sentient species, and all make an important contribution to sentience.

Hilgetag and Goulas[10] have extensively analyzed the various possible ways neural networks can be connected into diverse hierarchical arrangements:

> However, the sense in which 'hierarchy' is used in these publications can vary from one paper to the next, or even within the same paper. For example, when addressing '. . . the hierarchical arrangement of cortical sensory areas', a study may refer to concepts of laminar-specific projections, topological projection sequence, as well as a combination of both. Similarly, descriptions of how 'hierarchy' is expressed in the human and non-human primate brain may interchangeably employ different perspectives of 'hierarchy', such as distance along the posterior-anterior axis of the brain, ordered variations of neural responses in terms of functional complexity, gradients of cortical thickness, or a progression of laminar projection patterns. These are clearly very different matters, and while many neuroscientists have the intuitive feeling that these notions of 'hierarchy' are somehow related, it is impossible to establish whether this is true or not

and in which way the different interpretations may be linked without carefully exploring each of the different 'hierarchy' concepts in turn. Failure to do so is bound to result in confusion. (Hilgetag and Goulas, 2020, p. 1)

The authors make a very important point here with reference to neural hierarchies. From the standpoint of the emergence of sentience, it is critical to pay attention to the ways that these hierarchies operate, especially—as we shall see—when we consider the various ways that the scientific and philosophical literature has dealt with the idea of "levels" and the emergence of consciousness.

Levels, Hierarchies, and the Neurobiology of the Emergence of Sentience

For the analysis of neural hierarchies that are especially relevant to the neurobiological emergence of sentience, we can roughly divide the major subtypes of biological and especially neurobiological hierarchies into partially overlapping subtypes: *serial (nonnested) isomorphic (somatotopic) pathways* and *scalar (compositional) nested hierarchies*. While in principle these two can be distinguished, it is important to point out that both of these architectonic patterns may be and often are simultaneously present to varying degrees in any particular emergent neurobiological system.

Serial (nonnested) isomorphic pathways   As we discussed in chapter 3 regarding my proposed criteria for sentience, serially arranged isomorphic sensory pathways that are responsible for the feelings of vision, taste, olfaction, audition, and touch are the bread and butter of exterosensory awareness of the world. The hierarchical organization of these pathways display in part a nonnested organization in that the "higher levels" of the hierarchy are not *physically composed* of the lower levels, but rather the lower levels *project* their inputs to higher levels. In these nonnested organizations, the "levels" are therefore defined by their ordering in these synaptic pathways (e.g., primary, secondary, tertiary, etc.) that typically operate in an anatomically caudal (lower in the nervous system) to rostral (higher in the nervous system) pattern, or "upstream" (initial processing) or "downstream" (later processing) direction. A well-known example of this sort of organization is the retinotopic visual map (figure 8.2).

In a similar point-by-point fashion, the touch *somatotopy* of the surface of the skin is preserved at successive neural levels, so that the adjacent parts of the body are represented by adjacent neurons in the brain. In the *tonotopy* (tone mapping) of the auditory system, the vibrations of different

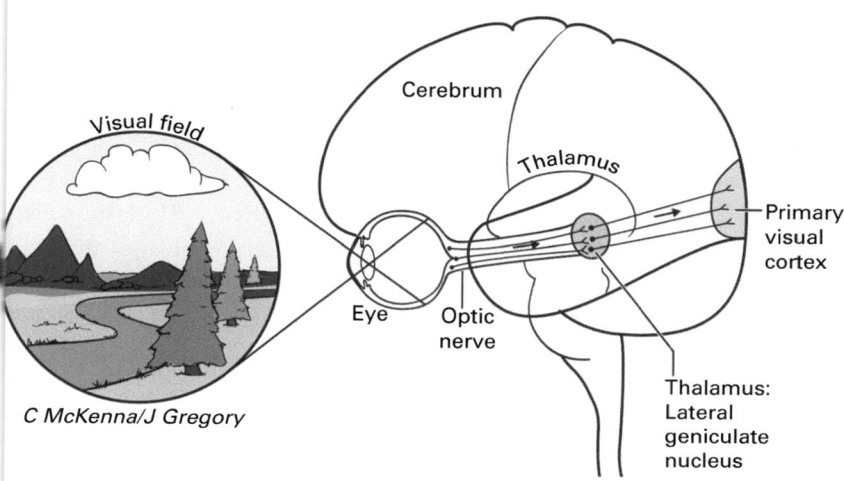

Figure 8.2

A simplified outline of the creation of the visual retinotopic maps and visual (sentient) images in the human. The visual pathway from the retina to primary visual cortex of the cerebrum is a *serial, multilevel, hierarchical nonnested pathway*. The mapping of the visual field as the information ascends through the visual hierarchy retain its isomorphic (retinotopic) organization. While each level of the hierarchy is synaptically connected to higher levels, note that the organization is *nonnested* since each level is physically independent (not contained *within*) higher or lower levels.

tones of sound are organized by their spatial positions on the cochlea, which is the receptor structure in the inner ear, and these mappings are in turn preserved at higher brain levels. So the creation of these detailed neurohierarchical maps are absolutely essential for the creation of this sort of exteroceptive sentience (figures 3.1 and 8.2).

But as was discussed in the introduction, all types of "feeling," whether from the world (exteroceptive) or from the body (interoceptive-affective; pain and pleasure, affect, etc.) are evidence of sentience or "something it is like to be." And as we have already seen, the underlying neuroanatomy of these various types of sentience are quite diverse. So while some researchers have considered the *sensory mental images* that result from exteroceptive sensory processing to be the prototypical manifestations of "consciousness," interceptive-affective "feeling" does not require the strictly nonnested serial somatotopy that we see in the creation of exteroceptive awareness. Additionally, the neural infrastructure of interoceptive-affective feeling also differs from the neural infrastructure of exteroceptive sentience in the subtype of neurons and the subparts of the neural infrastructure that are involved (figure 3.2). These differences will have important implications for the emergence of sentience and its hierarchical system nature that I address later on.

Scalar (compositional) nested hierarchies There is yet another way in which ES3 hierarchies promote the emergence of sentience. These are called "scalar (compositional) nested hierarchies" in which the ordering of "higher" and "lower" levels is determined by an *increase in the number of parts* (e.g., neurons) so that "higher" levels are greater in *scale* in comparison to "lower" levels." As scalar levels increase, there is also a simultaneous increase in the cellular differentiation of neurons at progressively higher levels.

In comparison to nonnested hierarchies, a distinguishing feature of nested or compositional hierarchies is that each higher level of the hierarchy is entirely *physically composed* of its constituent parts at lower levels and, conversely, the individual parts of the lower level are physically "nested" within the next higher level.[11]

A theoretical model that is based upon hierarchical scaling of brain connectivity is sometimes referred to as a "hierarchical modular network (HMN)." An HMN is organized in terms of progressive "modules within modules" in which smaller neural networks or "modules" are nested within larger neural modules at ever increasing scales. In terms of the pattern of emergence, we would anticipate that novel emergent features would arise as the scale of the global network progressively and hierarchically increases (figure 8.3).[12]

Hierarchical modularity is a complicated subject but Olaf Sporns,[13] an expert on the subject of neural computation and neural networks and an early advocate of HMNs, nicely summarizes the relationships and differences between nonnested, serial, topographical maps that we have just discussed and nested HMNs:

> Network hierarchy is often invoked in the context of modularity. Here, instead of a processing hierarchy, hierarchical levels are defined as levels of modularity, or nested arrangements of modules within modules. Systems with hierarchical modularity can be recursively subdivided into ever smaller and denser modules. The brain's anatomic organization, with individual neurons that are grouped into local populations, which are in turn grouped into brain regions and large-scale systems, resembles a network with hierarchical modularity. Hierarchical modularity has important implications for the temporal structure of brain dynamics. (Sporns, 2011, pp. 193–194)

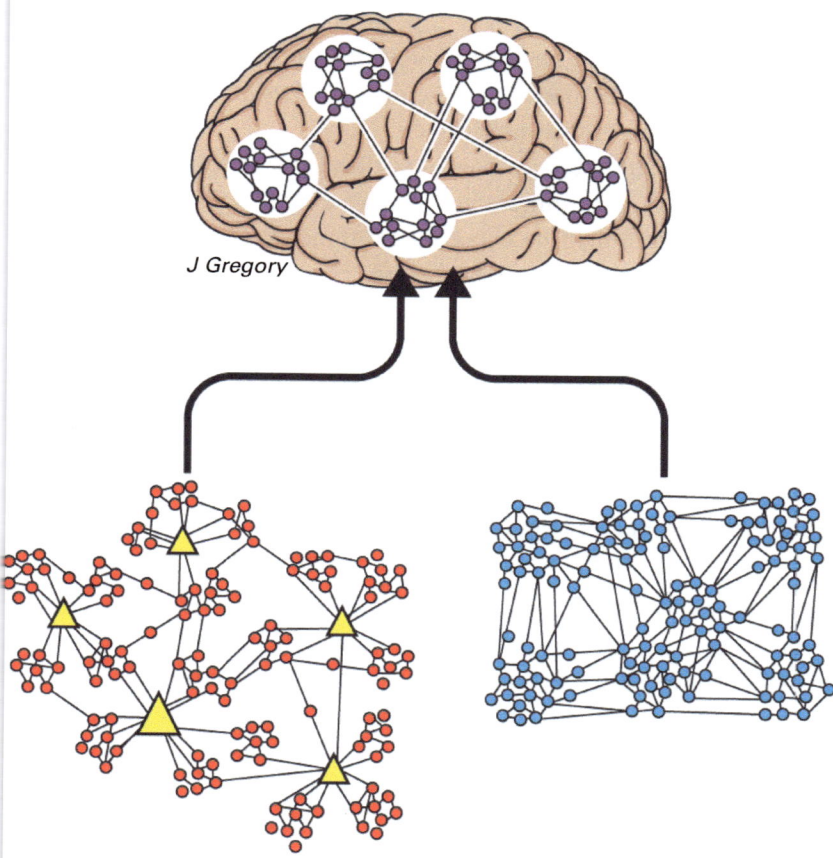

Figure 8.3

A schematic illustration of a hierarchical multilevel cortical network with both nested and nonnested architectures. A model that features hierarchically nested and nonnested brain connectivity is sometimes referred to as a "hierarchical modular network (HMN)" in which smaller neural networks or "modules" are nested within larger networks in increasing hierarchical *scales* (e.g., "modules within modules"). In theory, some of these modules may have central "hub nodes" (lower left, yellow triangles) that provide more centralized access at relatively lower scalar levels. In an HMN, novel features could emerge as the *hierarchical scale* as well as the *hierarchical level* of the global network progressively increases. (Based on Hilgetag and Goulas, 2020. See also Sporns, 2011.)

However, as shown in figure 8.3, some HMN models are not entirely nested and lower nested modules may be connected to higher modules in a nonnested (serial) fashion and, therefore, an HMN may show *both* nested and nonnested hierarchical arrangements.

Note that despite these various hierarchical patterns, there is no "top" where the parts of the entire system "come together" and where the emergence of sentience might be hypothesized to occur. Even if one supposed that the nervous systems "parts" are the "bottom" of the neural hierarchy, at the same time these "lower-level" elements *both create and interact with* "higher" aggregate levels in a "circular" causal pattern (see below) so it remains the case that any clear notion of "casual" neural effects of the parts and aggregate wholes is hard if not impossible to discern, especially given the diversity of emergent processes and their interactions within and across neural levels. We will return to this question of "emergent levels" in a moment since it plays a significant role in some misconceptions about the relationship between the emergence of sentience and neural hierarchies.

Temporal scales   In fact, the diverse mechanisms of emergent causality in neurobiologically complex brains is not even limited to serial connectivity or modular scale. The various neurohierarchical architectures that contribute to sentience are also operating at varying *temporal scales* that parallels their hierarchical patterns.[14]

Wolf Singer, who along with his colleagues were pioneers in the study of recurrent brain oscillations, made a similar observation.[15] During a symposium devoted to the question of "top-down" causation, a subject that we will consider in more detail later in this chapter and chapter 9, Singer described how there is what he called a "fantastic intertwining" of multiple hierarchical mechanisms that included wave frequencies, nestedness, and scales:

> I would like to make two points. One is on the nestedness of the organization of the brain. There is a fantastic intertwining of process of different scales, which are expressed in the wave frequencies characterizing our selections, for example, so that the higher the frequencies, the more local the process. It's a very close relation; we see these frequencies going from less than 0.1 Hz all the up to 30 Hz and more, and all this allows for a very, very complex network of relations; thus, nestedness is a very important aspect. (Singer, quoted in Auletta et al. 2013, p. 325).

Summary of the diversity of emergent mechanisms   Yet despite these variations in emergent mechanisms, all sentient species have advanced brain

mechanisms that include—at a minimum—caudal-rostral neurohierarchical organization, nonnested (serial) hierarchies, scalar-compositional
(micro-macro) levels, HMNs, and synchronized oscillations (table 7.1). In
addition to their diversity, these emergent mechanisms also occur across
multiple "levels of organization."

But here is the problem. Given the diversity and neurohierarchical complexity of these emergent mechanisms, unsurprisingly determining the
"levels" of emergence in a neurobiological system is problematic, yet this
issue is central for our understanding of the emergence of sentience. I consider this next.

Upward, Downward, and Circular Causality

So in contrast to viewing the relationship between the emergence and the
nervous system in terms of "layered levels," it is better to consider this
problem within the context of diverse *hierarchical directions* of their diverse
mechanisms. Jaegwon Kim in his analysis of the relationship between these
levels describes what he refers to as a *vertical directionality* that he considered
an "ordering relation that generates an "upward" direction and a "downward" direction of causality:

> We see that three kinds of inter- or intra-level causation are possible: (i) same-
> level causation, (ii) downward causation, and (iii) upward causation. Same-level
> causation, as the expression suggests, involves causal transactions between prop-
> erties at the same level—including cases in which one emergent property causes
> another emergent property to be instantiated. Downward causation occurs when
> a higher-level property, which may be an emergent property, causes the instantia-
> tion of a lower-level property; similarly, upward causation involved the causation
> of a higher level property by a lower-level property. (Kim, 2000, p. 309)

Thus, *upwardly causal* or "bottom-up" neural interactions are causal
interactions from lower to higher neurohierarchical "levels" of the system.
The nonnested serially organized isomorphic sensory pathways that rostrally converge upon "higher areas" of the neuroaxis are perhaps the clearest
example of a sentient subsystem that can be viewed as "bottom-up" emergent pathways that are involved in the creation of exteroceptive feelings. In
contrast, *downwardly causal* or "top-down" interactions are in the opposite
direction and refer to causal interactions that flow from higher to lower
neurohierarchical levels.[16]

While the emergence of sentience tends to be viewed in most neurohierarchical models as a "bottom-up" phenomenon, "top-down" causal interactions are also critical for the emergence of sentience.

But note, most importantly, that these terms *do not imply* that novel emergent features are "emerging" solely at a higher or lower level; but rather the direction of neural causal interactions may occur from lower to higher, higher to lower, or within the same levels of the neural axis, so it is more accurate to say that an emergent feature such as sentience is a *system feature, not a level feature*. These relationships are "heuristically" summarized in figure 8.4.[17]

Downward causation and constraint   While upward causal mechanisms is how most people typically think about the creation of emergent processes, less is said about downward causal interactions within the hierarchical emergent organizations that play a role in the creation of sentience.

For example, many standard *biological* models of hierarchical systems refer to downwardly ("top-down") causal effects as *constraints*. The term "constraint" has various meanings depending upon the context in which it is used. For example, in evolutionary biology theory, it refers to the limitations or biases to evolutionary change by natural selection.[18] But in biological systems theory, which is our concern here, *constraints* typically address

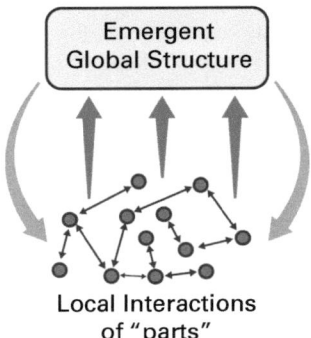

Figure 8.4
An illustration that emphasizes sentience as an emergent feature of a complex hierarchical system "as a whole." An approach to the problem of "layered levels" is to consider their causality in the context of their *hierarchical directions* (intra-level, upward, downward, and circular) that is created by an array of diverse neural hierarchical mechanisms. (Adapted from Lewin, 1992, p. 13.)

processes wherein higher levels in the hierarchy impose control over lower levels,[19] a relationship that in most respects is a form of biological downward causation that we discussed earlier.

The biologist H. H. Pattee was among the first to emphasize the critical role that constraint plays in the emergence of novel properties in biological systems. He also presciently predicted its role in operations of the brain:

> If there is to be any theory of general biology, it must explain the origin and operation (including the reliability and persistence) of the hierarchical constraints which harness matter to perform coherent functions. This is not just the problem of why certain amino acids are strung together to catalyze a specific reaction. The problem is universal and characteristic of all living matter. It occurs at every level of biological organization, from the molecule to the brain. It is the central problem of the origin of life, when aggregations of matter obeying only elementary physical laws first began to constrain individual molecules to a functional, collective behavior. It is the central problem of development where collections of cells control the growth or genetic expression of individual cells. It is the central problem of biological evolution in which groups of cells form larger and larger organizations by generating hierarchical constraints on subgroups. It is the central problem of the brain where there appears to be an unlimited possibility for new hierarchical levels of description. These are all problems of hierarchical organization. Theoretical biology must face this problem as fundamental, since hierarchical control is the essential and distinguishing characteristic of life. (Pattee, 1970, p. 119)

I have already described at least two ways in which downward causation can occur in neural hierarchical systems. First, there are anatomically rostral ("higher") to caudal ("lower") pathways—some of them reciprocal or synchronous—that function in a downwardly causal direction. Second, there is macro to micro scalar causality that occurs within nested hierarchical emergent system models. And some models such as HMN (figure 8.3) could combine both of these mechanisms.

But in addition to these, we must add that higher neurohierarchical levels and emergent processes also serve to *constrain* lower levels and, as Pattee puts it, "harness matter" to perform coherent functions. And note that downward constraints are present in *all* emergent aggregate biological systems. For NBE, it is especially important that Pattee opines: "It is the central problem of the brain where there appears to be an unlimited possibility for new hierarchical levels of description."

This is so because in neurohierarchical multilevel systems, the degree, diversity, and complexity of *both* downward causality and constraint and upward causality and emergence necessitate a broader view of these mechanisms that may be called "circular causality." Nunez provides the following regarding circular causality and scalar levels.[20]

> Brains are actually complex in the sense understood in the new field of complexity science. Small-scale complex systems often act bottom-up to generate emergent systems at larger scales. Such emergent systems exhibit novel properties and behaviors that are absent from their component parts. These larger systems then act top-down to influence smaller systems, thereby completing a loop of circular causality. (Nunez, 2016, p. 37)

From this we can see that it is not only via *upward neural interactions, causality, or emergence* that hierarchical systems create novel properties; it is also via *downward neural interactions, causality, or constraint* that hierarchical systems acquire emergent properties (figure 8.4). And the increased degree of *upward, downward, and circular interactions* in the brains of sentient animals geometrically magnifies the degree of the emergent features of the system as a whole.

Finally, it would appear that the upward, downward, and circular causality and their resulting emergent features are so intertwined, so *nondissociable* and mutually causal that the direction of causation as well as the varied notions of "levels of emergence" in the brain become at best ambiguous and at worst a heuristic abstraction.

Summary

In summary, there are numerous biological and especially neurobiological emergent mechanisms. Some of this diversity is specifically related to hierarchical patterns and interactions such as serial (nonnested), isomorphic (somatotopic) pathways, versus scalar (compositional) nested hierarchies, versus modular networks (HMN) that display aspects of both of the other two. In addition, there are other mechanisms of emergence that rely upon temporal integration based upon wave frequencies. And finally, there is also variability regarding the "direction" of the causation (e.g., "same level" versus "bottom-up" versus "top-down" causality and "constraint," versus "circular" causality). Therefore, sentience does not emerge at the "top" of the brain. Rather, it is an aggregate emergent system property.

I will return to the issue of the personal nature and character of sentience in chapter 10. But first, in the next chapter, I consider some other theories that relate to the issues addressed by NBE.

9   Neurobiological Emergentism: Panpsychism, Biopsychism, the "Emergentist Dilemma," and Integrated Information Theory

There are many theories of sentience and consciousness in general, some of which I have already described, that address how we might close the explanatory gap between the brain and experience. In this chapter, I focus on some theories that have a specific relationship to Neurobiological Emergentism (NBE). These include *panpsychism*, *biopsychism* and the "emergentist dilemma," and *Integrated Information Theory* (IIT). I argue that the first two of these create unnecessary barriers to the emergentist point of view that NBE solves; and that in the case of IIT, although this theory differs in many respects from NBE, IIT actually supports many of the emergentist positions of NBE, albeit without the proponents of IIT specifically recognizing this relationship between IIT and emergence.

Panpsychism

Panpsychism is a somewhat complicated set of theories that largely have in common the view that consciousness is a fundamental feature of the world. While its proponents are many and varied, there are some aspects that bear most directly on NBE.

Here is how Goff and colleagues, recently defined the meaning of the term:

> The word "panpsychism" literally means that everything has a mind. However, in contemporary debates it is generally understood as the view that mentality is fundamental and ubiquitous in the natural world. Thus, in conjunction with the widely held assumption (which will be reconsidered below) that fundamental things exist only at the micro-level, panpsychism entails that at least some kinds of micro-level entities have mentality, and that instances of those kinds are found in all things throughout the material universe. So whilst the panpsychist holds

that mentality is distributed throughout the natural world—in the sense that all material objects have *parts* with mental properties—she needn't hold that literally everything has a mind, e.g., she needn't hold that a rock has mental properties (just that the rock's fundamental parts do). (Goff et al., 2021, p. 5).

In terms of emergence theory, one might say that panpsychism is the *opposite* of an "emergence at the top" view. The panpsychist position holds that the basic required elements for "consciousness" are present even *below* the level of life. Here is how John Heil explains how panpsychism relates to emergentism:

> The idea, rather, is that conscious qualities are not emergent, not epiphenomenal, not add-ons. Conscious qualities 'go all the way down'. Electrons might not have minds, but, just as they possess charge and mass, electrons possess conscious, experiential qualities. There is something it is like to be an electron. The brain might appear to be soggy gray matter, but its component parts exhibit flickers of consciousness. The brain has a Technicolor phenomenology because its parts do. (Heil, 2019, p. 228)

One of the significant motivators for the panpsychist position is that the mechanisms of the emergence of sentience from the brain remain—in some writers' opinions—completely mysterious. Indeed, in Goff et al.'s review of panpsychism, they discuss what they called the panpsychist "anti-emergence argument" in which they attribute one of the motivators for the panpsychist position to the "failure" of other theories to explain how *material* neurons create *subjective* experiences which appears to them to be beyond standard emergent principles:

> There is no reason to suppose that "further scientific investigation" has to be pursued under the methodological assumption that *consciousness is to be accounted for in terms of processes which don't involve consciousness,* e.g., in terms of facts about non-conscious neurons. The panpsychist proposes an alternative approach: *explain human and animal consciousness in terms of more basic forms of consciousness.* These more basic forms of consciousness are then postulated as properties of the fundamental constituents of the material world, perhaps of quarks and electrons. Thus, we try to explain the consciousness of the human brain in terms of the consciousness of its most fundamental parts. (Goff et al., 2021, p. 9)

Thus, many panpsychist theories have in common an effort to reconcile the *physical* aspects of matter—including neurons and brains—with the *subjective* "what it is like to be" aspects of sentience.

But as I have argued, this tension between materialism and the emergence of sentience is unnecessary. This is because the reason for this apparent dichotomy between the "physical" and the "mental" (sentience) is simply the result of the failure to appreciate the progressive biological and evolutionary increase in standard emergent features of brains as they advance toward sentience; and just how unique and important this monumental increase in emergent processes is.

But as noted earlier, there are two major factors that obscure this insight; first, the exponential increase in emergent properties and its aggregate emergent effects make it *virtually impossible* to fully trace its biological and neurobiological mechanisms. This we have already discussed. The other issue is its *personal subjective nature*. I explore this further in the next chapter.

Biopsychism and Reber's "Emergentist's Dilemma"

The term "biopsychism" was coined by Ernst Haekel in 1892 and it is now conventionally come to mean the view that all living organisms are sentient.[1] So what are the implications of this for the emergentist view?

One of Reber's principle arguments for his CBC model that was discussed in chapter 5 is that sentience could not have *naturally emerged* somewhere between a bacterium and sentient animals and therefore sentience must be present in bacteria in the first place. This is a variety of explanations of sentience as a "lower order" feature similar to panpsychism with the difference that it pushes the consciousness level from *all matter* to just *living matter*. In Reber's case, he refers to this as the *Emergentist's Dilemma*, and it is perhaps the biggest fulcrum upon which he bases his claim for single-celled consciousness:

> In addition to creating many a dispute among researchers about just how these data are interpreted and what "counts" truly and unambiguously, as evidence for consciousness, there's an even bigger problem—the one I'll be calling the *Emergentist's Dilemma*. It's not simply a matter of anointing a species (or clade or phylum) with an ontologically real consciousness; a coherent argument must be developed that shows why, when evolutionary mechanisms finally produced a species with sentience, consciousness this mind-like thing burst into existence where one cosmic moment earlier all species of lesser complexity were mere mechanical entities behaving without a glimmer of awareness. (Reber, 2019, p. xviii)

This challenge to emergentism can be explained by several points. First, as argued in chapter 5, these basic ES1 organisms are not sentient. Michael Woodruff provides an excellent summary of why the proposal of unicellular consciousness is not supported. He also notes that *sensing* and its associated behaviors in these organisms can be fully explained by the molecular, biological, chemotaxis mechanisms and do not require additional explanation nor do they prove the presence of sentience:

> So we have detailed knowledge concerning the genes behind the construction of the sensory and motor substrates underlying chemotaxis in bacteria. The biochemical basis of both the extracellular and intracellular transduction processes that cause the flagellum to rotate in one direction or the other are also understood. This information seems to be quite sufficient to explain bacterial chemotaxis without the introduction of sentience as an additional explanatory process.
>
> In sum, then, CBC theory fails at the level of two of its fundamental assumptions. First, there seems to be no empirical reason to set aside Occam's infamous razor and admit sentience as an explanatory variable to explain bacterial chemotaxis. Second, there is no good evidence that evolutionary continuity in the genetic substrates of bio-sensitivity exists between bacteria and organisms with nervous systems. Therefore, despite Reber's ambitious program, CBC does not supplant the more traditional view that a nervous system is necessary for sentience. (Woodruff, 2016, p. 2)

Second, consistent with the evolutionary origins of sentience, life processes play a basal and fundamental role in enabling the eventual emergence of sentience. So while I do not endorse the view that all life is sentient, my analysis of the step-wise emergence of sentience begins with basic life processes and single-celled organisms that in my view are necessary first steps for the long (3.5 billion years) evolutionary emergence of sentience.

So NBE has no emergentist's dilemma; in fact, emergence holds the key to understanding how sentience is created and evolved. Applying the principles of biological emergence that are outlined in table 2.1 to unicellular organisms, we find no emergentist's dilemma but rather an *emergentist's solution;* and that all the biological principles that are involved in the emergence of sensing and their associated behaviors in single-celled organisms (table 5.1) are consistent with the principles of biological emergence in general.

Thus, I propose that while *sensing* is an evolutionarily early emergent feature of life, it does not constitute *sentience*. However, it is wholly possible, reasonable, and logical that sentience is an emergent feature of *both* sensing and life.

Phi and Integrated Information Theory

IIT is a theory of consciousness originally proposed by Giulio Tononi and later developed along with colleagues. Many aspects of IIT are complex and entail numerous axioms, postulates, and some mathematical formulations that are largely aimed at explaining how consciousness is created. A full exploration of the theory is beyond my current concerns and the reader is referred to some sources.[2]

That said, despite its apparent overall complexity, the central hypothesis of IIT includes some aspects of the theory that are quite relevant to NBE, and on this point, the theory is relatively straightforward.

First, here is how Tononi describes IIT:

> Integrated information theory (IIT) attempts to identify the essential properties of consciousness (*axioms*) and, from there, infers the properties of physical systems that can account for it (*postulates*). Based on the postulates, it permits in principle to derive, for any particular system of elements in a state, whether it has consciousness, how much, and which particular experience it is having. IIT offers a parsimonious explanation for empirical evidence, makes testable predictions, and permits inferences and extrapolations. (Tononi, 2015, p. 1)

At the center of IIT is the concept of "phi" or Φ that Tononi describes as a measure as "integrated information:"

> The cause-effect structure specified by the system must be *unified*: it must be intrinsically *irreducible* to that specified by non-interdependent sub-systems obtained by unidirectional partition . . . Intrinsic irreducibility can be measured as integrated information ("big phi" or Φ, a non-negative number), which quantifies to what extent the cause-effect structure specified by a system's elements changes if the system is partitioned (cut or reduced) along its minimum partition (the one that makes the least difference). By contrast, if a partition of the system makes no difference to its cause-effect structure, then the whole is reducible to those parts. If a whole has no cause-effect power above and beyond its parts, then there is no point in assuming that the whole exists in and of itself: thus, having irreducible cause-effect power is a further prerequisite for existence. This postulate also applies to individual mechanisms: a subset of elements can contribute a specific aspect of experience only if their combined cause-effect repertoire is irreducible by a minimum partition of the mechanism ("small phi" or φ). (Tononi, 2015, p. 4)

There are some significant common features and differences between NBE and IIT. First, we discuss their commonalities, and some are quite surprising.

Common Features between NBE and IIT

Emergence, Phi, and "Integrated Information"

One of the most striking aspects of the concept of phi is how strongly it invokes *standard emergent principles* even if the word "emergence" isn't used.

Thus, according to the above quote, a system with phi must be *unified, irreducible to its parts,* and the *whole must have cause-effects power above and beyond its parts* and that a subset of elements can make under some conditions a *causal contribution to a specific aspect of experience.*

In fact, these features of phi are undeniably all *basic features of emergence.* Indeed, in the light of the theory of NBE, one could convincingly argue that phi is actually one—albeit mathematically oriented—measure of the *emergence* of consciousness.[3]

Christof Koch, a proponent of IIT, like Tononi affirms the relationship between emergent principles and consciousness when he points out that phi measures "the extent to which the system is more than the sum of its parts."

> The parts—the modules—of the system account for as much non-integrated, independent information as possible. Thus, if all of the individual chunks of the brain taken in isolation already account for much of the information, little further integration has occurred. Φ measures how much the network, in its current state, is synergistic, the extent to which the system is more than the sum of its parts. Thus, Φ can also be considered to be a measure of the holism of the network, (Koch, 2012, p. 127)

It hard to imagine more direct statements that phi is—at least in some part—a measure of the degree of emergence of a neurobiologically complex integrated system.

The Role of Hierarchical Interactions

A second common feature between NBE and IIT is that both theories emphasize the importance of the recurrent, reciprocal, and "circular" interactions between the parts and the neural "levels" (however they are distinguished) that we discussed earlier. But this point is not unique to either theory; indeed, this factor is featured with varying emphasis in many neurobiological theories of consciousness.[4]

Differences between NBE and IIT

Despite these similarities, there are also some significant points of disagreement between NBE and IIT regarding the *relationships* between emergence, consciousness, and sentience. Here is where they differ.

Compared to NBE, IIT Claims a Very Low Degree of Complexity for the Emergence of Consciousness

IIT proposes an extremely low degree of emergence or complexity that is required for the creation of consciousness or sentience. Indeed, it is so low that it has been considered by some as endorsing panpsychism.[5] The proponents of IIT have argued that objects without interacting parts are not conscious nor are the parts of any conscious system whose integrated information is less than φMax[6] However, IIT does claim that anything with *interacting parts* can have consciousness and a phi value above 0. And this then includes simple, nonliving entities such as thermostats or an isolated proton with its three interacting quarks.[7] In any event, the bar is very low when compared to NBE.

Phi Is Continuous Whereas NBE Is Punctuated

IIT posits its central factor phi as a *continuous* function of the degree of "integrated information" that increases along with the increasing complexity of the system in question. In contrast, NBE holds that sentience progressively emerges in *punctuated stages* from nonsentient life and does clearly emerge until approximately 520 million years ago.

I have already offered the evidence for this hypothesis. Therefore, all living organisms, despite having all manner of emergent properties with values of phi far beyond 0, are not sentient. Thus, either all these biologically and neurobiologically relatively simpler animals including single-celled organisms are sentient—and I have marshalled evidence for why this position is incorrect—or IIT is incorrect. The NBE position is that it is the *degree* of emergence that ultimately determines the emergence of sentience, not the mere presence of any degree of emergence or "integrated information."

The Question of the Relationship between Life, Biology, and Sentience: Why Substrate Matters for the Creation of Sentience as "Feeling"

The argument for the NBE claims that *substrate matters* hinges on two points. First, the obvious one, that the NBE model pertains to *animal sentience,* and that animal sentience is an emergent feature of life. As such, according to in this view, all the features of the emergence of life are still operative throughout the progression of stages that lead to sentience. This much is easy to explain.

A second related but more difficult question is what is called the problem of the "character of consciousness" that was introduced in chapter 1. This is the question of why so-called "qualia"—feelings" of all type—feel the *particular way* that they do. For instance, why does "red" feel different from "blue" or "happy?" This is what Chalmers calls the problem of *the character of consciousness* that was discussed in chapter 1.

On this point, Tononi's IIT explanation is that each and every quale has a distinct "conceptual structure" in a many-dimensioned graph of "elements" in a state—a "form" in cause-effect space. Here is how he describes it:

> Together, the axioms and postulates of IIT provide a principled way to determine whether a set of elements in a state specifies a conceptual structure and, if so, to characterize it in every aspect. The central identity proposed by IIT is then as follows: every experience is *identical* with a conceptual structure that is maximally irreducible intrinsically, also called "quale" sensu lato . . . In other words, an experience *is* a "form" in cause-effect space. The quality of the experience—the way it feels due to its particular content of phenomenal distinctions—is completely specified by the form of the conceptual structure: the phenomenal distinctions are given by the concepts (qualia sensu stricto) and their relationship in cause-effect space. The quantity of the experience—the level to which it exists—is given by its irreducibility Φmax. (Tononi, 2015, p. 4)

What a "conceptual structure" is to say the least a conceptually abstract postulate. And note again that this explanation makes no reference to the biology of the different sentient experiences that nervous systems can create.

In my view, it is much more logical and straightforward to simply propose that all qualitative experiences—both within (e.g., "red" versus "blue") and across (exteroceptive versus affective) categories—while they have many emergent principles *in common* beginning with the emergent biology of life, they also clearly and unambiguously *differ* in their neurobiological substrates. But IIT claims that the actual physical substrate that creates sentience is irrelevant to both its creation and the manner in which it is experienced. Here is how Ellia and Chis-Ciure in a recent comparison of Feinberg and Mallatt's earlier theory of NN and IIT expressed it: "IIT allows consciousness to be realized in multiple substrates, in fact the substrate per se is irrelevant, as long as it specifies the integrated information described by the mathematical formalism of the theory" (Ellia and Chis-Ciure, 2022, p. 5).

This is a complicated point regarding how NBE differs from IIT. On the one hand, I have already enumerated the various ways that sentience is realized in somewhat different substrates, for instance exterosensory versus affective sentience. But according to NBE, and in contrast to IIT, despite this degree of diversity in the emergence of sentience, its biological and neurobiological substrates do matter.

And these varying neural architectures run all the way from basic *sensing to sentience*. In other words, the neural "parts" of each system and the emergent system properties for each "quale"—from sensory receptors and affective centers all the way to more globally scalar emergent "levels"—are ultimately and markedly *materially different* and they vary significantly accordingly to their accompanying experiential features. In other words, *the parts and their interactions are causally related to the qualia they create.* Otherwise it is hard, if not impossible, to explain the personal "character of experience."

I will have more to say on this point in the next chapter, where we consider the final question: *the personal nature of sentience.*

Because mental phenomena are essentially connected with consciousness, and because consciousness is essentially subjective, it follows that the ontology of the mental is essentially a first-person ontology. Mental states are always somebody's mental states. There is always a "first-person," an "I," that has these mental states.
—John Searle, 1992, p. 20.

How Neurobiological Emergentism Explains the Personal Nature and Character of Sentience

As discussed in chapter 1, Levine's view and that of some philosophers of consciousness[1] is that the "gap" between the objective features of the brain versus the subjective and personal character of consciousness is what makes sentience so perplexing and mysterious. It is also what makes theories such as strong emergence seem like a reasonable—if not a default—position.

So how can the subjective features of consciousness be explained by objective neurobiological science? How can the gap be closed? Here is the Neurobiological Emergentism (NBE) explanation of the natural emergence of the *personal nature and character* of sentience.

The Emergence of Sentience Is Neurobiologically Complex but Not Fundamentally Different from Biological Emergence in General

First, as I have already emphasized, NBE postulates that the explanation of the emergence of sentience does not require some radically unique *type* of emergence (e.g., physically fundamental, "strong," "at the top," or dualistic, panpsychist, or biopsychist) that makes sentience possible; rather it is the *degree* of standard biological emergence that is made possible by the evolution of complex and hierarchical nervous systems (chapter 4) that enables the emergence of a novel system feature such as sentience.

The Personal Nature of Sentience Is an Outgrowth of Its Emergent Biological Basis

Second, once the first problem is understood and appreciated, it becomes readily apparent and "nonmysterious" how sentience can have a personal existence or ontology. First, because consciousness is built upon the emergence of life in any single organism, and because both life and consciousness are system features of embodied organisms, then it follows that sentient feelings (perceptions, "qualia," etc.) are emergent system features of certain neurobiologically evolved brains, and each feeling is a personal system feature of that individual living organism just as life itself is an embodied personal system feature of the organism. Therefore, life provides the *natural initial conditions* for the emergence of a personal sentience.[2] In short, life entails embodiment that ultimately allows and indeed *necessitates* the individual personal nature of sentience. I will say more about this in chapter 11.

These Two Variables of the Neurobiology and Personal Nature of Sentience Coemerge and Coevolved

Third, and perhaps most importantly, if *we put together the first two postulates,* we can see how the emergence of sentience and its personal nature both biologically cooccur and evolutionarily coevolved.

We begin with life as an emergent feature of standard biological processes that entails basic sensing capabilities as shown by ES1 microorganisms. Then, via evolution, additional neurobiological features are added to the system in ES2 animals that create more advanced presentient capabilities. And when these emergent mechanisms reach a certain stage of neurobiological and neurohierarchical complexity and novelty, there results the emergence of the various forms of sentience as unique aggregate system features in ES3 animals.

But what is also essential to appreciate is that as these novel emergent processes are advancing and evolving, so too pari passu is its personal nature as a *system feature of the embodied organism.* Thus, NBE explains how *both* sentience and its personal nature are naturally emergent aggregate system features of complex brains.

Why This Is Important for the Science and Philosophy of Sentience

So I believe that we are closing in on the "gap" between the physical brain and sentience and consciousness. But there are still a few more things that we need to explain about the personal nature of sentience if the explanatory "gap" is to be completely eliminated.

Some of the views that invoke emergence in an attempt to scientifically close the gap between the brain and personal experience entail some of the theoretical positions that I have discussed, including positions that advocate the "strong" emergence of consciousness. And many of the questions that were raised by philosophers such as Broad and Levine regarding the "explanatory gap" can be traced to their perplexity regarding how something like a uniquely personal sentience can emerge from the brain and have its particular subjective "character."

For this reason, some theorists (especially philosophers) were led to endorsing some variety of mind-body "dualism;" but as I have argued, by elucidating the personal nature of sentience and consciousness, we can more naturally explain these enduring enigmas.

Drawing upon some of the issues that I discussed earlier, I think these theories are basically trying to resolve two related problems. The first was discussed in chapter 8. This is the appearance or assumption that the *personal experiential qualities* of sentience somehow emerge from the "highest" neural levels of the brain, creating the appearance that these sentient qualities in some sense emerge "separate" from the brain. It then it could be supposed that these experiential—possibly "nonmaterial"—qualities could have *causal effects* upon the "lower-level" material brain.

The second problem that they struggle with is how can we reconcile the obvious *experiential differences* between the brain as observed or explained (third-person perspective) versus the brain as directly experienced (first-person "something it is like to be" perspective). This problem was described in chapter 1 and I address it here.

The Personal Nature of Sentience and the "Experiential Gap"

While I find no *scientific* explanatory gap between the brain and the emergence of subjective personal experience, there remains an *experiential gap* between objective scientific explanations of the brain and subjective experience.[3]

On the one hand, I fully agree that no amount of knowledge about or description of brain functions can be equated with, fully capture, or can substitute for "something it is like to be"—the third-person versus first-person aspects of experience—any more than *describing* one's first-person experience can substitute for *having* that experience.

But armed with a credible naturalistic theory of *both* the emergence of sentience and its personal nature, we also can scientifically reconcile the experiential gap between these first-person and third-person points of view without invoking any dualism between the brain and the mind. In other words, NBE is entirely compatible with physicalism.[4] And I argue that NBE makes this experiential gap scientifically unproblematic. I explain this claim next (see figure 10.1).

And Then Along Comes Mary: The Distinction between an Explanatory Gap and an Experiential Gap

The *experiential gap* is nicely illustrated by a famous thought experiment proposed by philosopher Frank Jackson in an article entitled "Epiphenomenal Qualia"(1982) that he later expanded upon in a second paper entitled "What Mary Didn't Know" (1986). Over the years, it has come to be known as the "knowledge argument."[5]

The basis of Jackson's theoretical scenario is to imagine that there is a brilliant neuroscientist named Mary, who knows everything there is to know about the neurobiological physical properties behind color vision. Now the twist in the story is that, for whatever reason, Mary has been raised her whole life in a black-and-white room and therefore has never personally experienced color. The knowledge she lacks because of this is sometimes called "knowledge by acquaintance."[6] Then, as the story goes, Mary is finally released from her room and—low and behold—she *sees* color for the first time. The question is: What just happened to Mary?

In actuality, the argument is essentially the same one that was raised by Broad's archangel.

Both the archangel and Mary knew all the facts about the brain yet would be unable to deduce from those facts what color actually *feels* like. So the argument goes that she gains some new *knowledge* about the world—hence the *knowledge argument.*

There are many interpretations and debates about Mary and what she tells us about sentience. These are many and diverse and will not be reviewed here.[7] For my purposes, I will make one point regarding NBE and Mary.

While Jackson's actual conclusions about Mary are subject to debate, it is commonly taken to be an argument against physicalism; that is if personal sensory experiences cannot be *equated* with material brain states, then this creates a philosophical dilemma that, for some, leads to an argument against physicalism.

Our reason for bringing up Mary is that in essence it is the same issue that is raised by Levine, Broad, and Chalmers that we discussed in chapter 1. Here is how Jackson somewhat later expressed his original thoughts about what the Mary experiment means:

> The question is especially pressing in the case of the pangs, the pains, and the sensings of red. On the face of it, no amount of aggregation of elements with the kinds of properties the physical sciences talk of can make up the phenomenal, conscious side of our psychology. That side of things is, in one way or another, an extra; a part of the account of what our world is like that is left out by the physical sciences—or so it seems. (Jackson, 2004, p. xv)

However, later on, Jackson completely reverses his initial position and comes to a radically different conclusion to the effect that the Mary experiment does not support an argument against *physicalism*. But of particular importance for NBE is *why* he changed his mind:

> I now think that what is, on the face of it, true is, on reflection, false. I now think that we have no choice but the embrace some version or other of physicalism. We are nothing over and above a very complex aggregation of purely physical elements interacting with, and carrying information about, a vast complex of purely physical elements. And as I have no intention of embracing eliminativism about phenomenal consciousness . . . I think that, somewhere or other, in the vast, complex, purely physical aggregation that makes up our world, there is phenomenal consciousness. (Jackson, 2004, p. xvi)

Notice that in the first quote that reflects his initial views, Jackson denies that an *aggregation of elements* could produce "phenomenal conscious;" but in the second quote, many years later, he arrives at a very different opinion that indeed a *complex aggregation of interacting physical elements*—the very claim of NBE—could produce subjective experience. This carries with it the assumption that there is an *experiential* difference between *knowing* facts about experience and actually *having* an experience but this difference does not preclude a naturally emergentist explanation of how sentience is created by neurobiologically complex brains (figure 10.1).

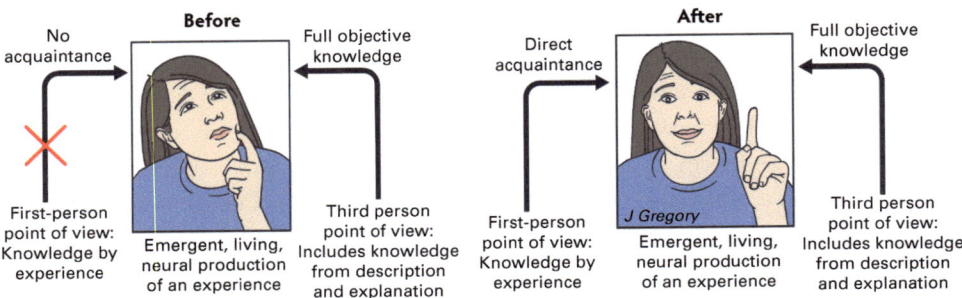

Figure 10.1
Mary and the "knowledge argument." Some kinds of knowledge can only be obtained by experience. "Knowing" is of two types: *experiential* via direct acquaintance or by *objective* descriptive or explanation. Before Mary has direct experience of color (left), although she has full objective knowledge of brain process, she lacks the first-person experience that she gains after she experiences color (right). The difference between "before" and "after" can simply be explained by the fact that only by *being* a living neural brain that produces the emergent process of experience can knowledge by acquaintance be acquired. Note that this view does not entail any *scientific dualism* between the brain and the mind, nor does it require a "nonphysical" explanation for sentience.

Conclusions

While there surely are some philosophical questions that remain, in my view these arguments—as Jackson ultimately acknowledged—do not bear upon the scientific question of how the brain creates subjective experience. NBE accounts for both the *natural emergence of sentience* as well as its *personal nature*; and that the latter derives in part from its *living biological embodiment* that can be traced back to its distant nonsentient origins from the first sensing organism. And thus, the *experiential gap* between scientific explanation and personal experience (figure 10.1) does not violate physicalism, nor does it support the various "strong emergence" theories of consciousness.

Thus, I believe that NBE has a persuasive argument against other theories that claim that sentience is imminent in nearly all matter (panpsychism), all life including single-celled organisms (biopsychism), all integrated matter (IIT), or that it's due to some fundamental novel physical forces of nature (naturalistic dualism; strong emergentism). Rather, we can explain the biology and neurobiology of sentience *and* its personal nature by the demonstration that they are both natural and novel *emergent features* of neurobiologically complex and neurobiologically evolved brains.

11 Neurobiological Emergentism: Putting It All Together

In this book I have discussed a range of issues that deal with various related aspects of the problem of understanding the neurobiological and evolutionary emergence of sentience. In this summary chapter, I return to the core principles of Neurobiological Emergentism (NBE) that I briefly introduced in chapter 1 and summarize where I propose four core postulates regarding how NBE integrates and unifies these approaches to explaining sentience.

Postulate 1: Sentience Is Ultimately an Emergent Feature of Embodied Life

I have proposed here and elsewhere that the neurobiological mechanisms that create sentience are emergent features of the embodied life of the organism. Therefore, any theory of the nature of the creation of at least all animal sentience must begin with life.

While I believe that the evidence supports the view that more basal ES1 organisms and animals at ES2 are not sentient, they clearly possess basic *sensing* capabilities; and these capabilities—like sentience—are emergent features of the animal's embodied life processes that serve as *basal system features* for the emergence of *both* sensing and sentience.

The role that the features of life play in the creation and emergence of consciousness is supported by some philosophers who are interested in explaining consciousness. In particular, Evan Thompson in his book *Mind in Life* has argued that the failure to appreciate the relationship between life processes and consciousness mistakenly draws an "unbridgeable divide" (an "explanatory gap") between the physical brain on the one hand and experience (feeling) on the other, a position that in essence disregards all the general biological features that can help explain consciousness and how its subjective features are created:

> I have argued that the standard formulation of the hard problem is embedded in the Cartesian framework of the "mental" versus the "physical," and that this framework should be given up in favor of an approach centered on the notion of life or living being. Although the explanatory gap does not go away when we adopt this approach, it does take on a different character. The guiding issue is no longer the contrived one of whether a subjectivist concept of consciousness can be derived from an objectivist concept of the body. Rather, the guiding issue is to understand the emergence of living subjectivity from living being, where living being is understood as already possessed of an interiority that escapes the objectivist picture of nature. (Thompson, 2007, p. 236)

Note especially that Thompson also highlights the relationship between the "emergence of living subjectivity from living being," as well as embodiment ("interiority") that "escapes the objectivist picture ("Cartesian framework") of nature." I believe that Thompson's philosophical point of view is consistent with and lends solid support for the view that I have presented here: That embodied life ultimately gives sentience its naturally evolving personal nature (chapter 10 and postulate 4 below).

But as Thompson notes, the explanatory gap by no means goes away simply because consciousness or sentience is a feature of embodied life. The role of life helps in part to close the "explanatory gap," but as Thompson suggests, something more is needed. I propose that the "something more" are the unique emergent neurobiological features of sentient animals that I have reviewed and summarize next in postulates 2–4.

Postulate 2: Sentience Evolved in Progressive Stages That Are Characterized by an Increase in Novel Emergent Features

I propose that the evolutionary stages of the emergence of sensing and sentience can be roughly divided into three biological and neurobiological emergent stages. I find no *single* biological feature that accounts for the emergence of sentience, but rather there is a progressive evolution of sentience in which there are *punctuated transitions* between the three emergent stages. And as noted earlier, it also appears that the time frame between these levels shrink as the evolution toward sentience proceeds (figure 4.1).

When we trace the emergence of sentience at ES3, the role that neurobiologically and hierarchically complex brains is an obvious feature of sentient brains (tables 4.1 and 7.1). Thus, when we compare ES1 and ES2 organisms to ES3 animals, there is what might be considered an "explosion" in some important neurobiological features that contribute to sentience. These features include a large increase in the *number* of neurons in the nervous system overall, the degree of the *specialized functions* of these neurons, the number of *hierarchical levels* in the nervous system, and the degree of the *interaction* of these levels (table 4.1).

This leads to a main point regarding my proposal that sentience is an emergent system feature of complex evolved brains: Note how the above features of nervous systems that create sentience at the ES3 stage compare with the generation of emergent features *in all emergent systems and emergent biological systems*. These general features (table 2.1) include novel *aggregate system properties* that are created by the dynamic interaction of the system's parts that requires that the system's parts to be *physically united, integrated, or at a minimum interacting in some fashion*. And, importantly, *hierarchical systems* are critical to an increase in the novel emergent system properties. Thus, the most general enumeration of the features of nervous systems that create sentience (table 4.1) is consistent with the factors that create novelty in emergent systems in general (table 2.1).

In conclusion, I offer that it should come as no surprise that if I am correct that sentience is a novel emergent property of complex neurohierarchical nervous systems that there is such a striking correspondence between the principles of emergence in general and the properties of nervous systems in animals that the evidence supports are sentient. This leads me to the next point.

Postulate 3: The Increasing Degree of Standard Neurobiological Emergence Is What Makes Sentience Possible

Recall from the Preface and chapter 8 that I discussed the issue of whether proposing some "strong" or "radical" form of emergence was required to explain the emergence of sentience; that sentience and consciousness must be the result of some "fundamental" physical property, a claim that I disputed. Chalmers expresses one version of this view:

> I suggest that a theory of consciousness should take experience as fundamental. We know that a theory of consciousness requires the addition of something fundamental to our ontology, as everything in physical theory is compatible with the

absence of consciousness. We might add some entirely new nonphysical feature, from which experience can be derived, but it is hard to see what such a feature would be like. More likely, we will take experience itself as a fundamental feature of the world, alongside mass, charge, and space-time. If we take experience as fundamental, then we can go about the business of constructing a theory of experience. (Chalmers, 1995, p. 210)

But in contrast to Chalmers' view—which is not entirely implausible but would be difficult to *prove* unless some "fundamental feature of the world, alongside mass, charge, and space-time" is actually identified—I propose that if my model is correct, that there are punctuated but natural and continuous transitions from *ES1→ES2→ES3,* then standard and natural biological emergent processes are scientifically sufficient to explain the emergence of sentience from the brain, just like all complex physical systems display emergent system properties that are the aggregate result of the parts of the system and their interactions. In other words, I find that there is an increasing *degree* or *magnitude* and *variety* of biological and neurobiological process (table 4.1) that account for the emergence of sentience, and thus there is no need to posit some fundamentally different *kind* of "strong" emergence for the creation of sentience.

In summary, I propose that the aforementioned correspondences provide solid evidence that it is an increase in the degree of standard emergent processes, as measured by the increase in the number and complexity of the neurobiological emergent processes—made possible in part by the evolution of hierarchical nervous systems—that accounts for the emergence of sentience; and while sentience is a novel emergent property of evolutionarily complex brains, the emergence of sentience does not require a *novel fundamental physics.*

Postulate 4: The Emergent Mechanisms of Sentience Are Diverse

I have proposed (chapter 8) that there are numerous biological and especially neurobiological mechanisms that contribute to the emergence of sentience. Some of this diversity is specifically related to hierarchical patterns and interactions such as serial (nonnested), isomorphic (somatotopic) pathways, versus scalar (compositional) nested hierarchies, versus modular networks (HMN) that display aspects of both of the other two. In addition, there are other mechanisms of emergence that rely upon temporal integration based upon wave frequencies. And finally, there is also variability regarding the "direction" of the causation (e.g., "same level" versus "bottom-up" versus "top-down" causality and "constraint," versus "circular" causality). Therefore, sentience does not emerge at the "top" of the brain. Rather, it is an aggregate emergent system property.

So, consistent with its emergent nature, there is a diversity of neurobiological mechanisms within subtypes of sentient experiences (exteroceptive versus interoceptive-affective; chapter 3) and across sentient animals (chapter 7). And just as subtypes of sentience are *experienced* differently, we may also assume it is at least plausible that sentience is *experienced differently* across species, despite the fact that the evidence seems clear that all ES3 animals are experiencing "something it is like to be."[1]

Postulate 5: NBE Can Naturally Explain and Reconcile the "Explanatory Gap"

So what about the "explanatory gap?" I think that, in general, the problem of the explanatory gap is how can we bridge the apparent gap between *objective* neural properties—many of which we scientifically know—to the *subjective* qualities of personal experience. This is indeed a "hard problem," but I think it is here that there is an opportunity to reconcile the science behind the creation of sentient states and the philosophical questions that these experiential states pose.

Let's start with Broad's archangel. Recall from chapter 1 that the central point that Broad raises in this scenario is that it appeared that no *knowledge* possessed by the scientifically omniscient archangel of the physical structure of both ammonia and the brain could substitute for the actual *personal experience* of "what it is like" to smell ammonia.

Levine was the first to refer to this as the "explanatory gap." As he put it "For there appears to be nothing about C-fiber firing which makes it naturally 'fit' the phenomenal properties of pain, any more than it would fit some other set of phenomenal properties. . . . One might say, it makes the way pain feels into merely brute fact" (Levine, 1983, p. 357).

The gap is also raised by the "Mary problem." Without a *personal experience* of color, even given all the facts of science of color perception, Mary could never know what the color red "feels" like.

Similarly, Chalmers raises the question of the "character of consciousness." He wonders how we can relate the objective facts of the science of feelings with the subjective experience they create:

When someone strikes middle C on the piano, a complex chain of events is set into place. Sound vibrates in the air and a wave travels to my ear. The wave is processed and analyzed into frequencies inside the ear, and a signal is sent to the auditory cortex . . . But why should this be accompanied by an experience? And why, in particular, should it be accompanied by that experience, with its characteristic rich tone and timbre? (Chalmers, 1996, p. 5).

A Two-Part Emergentist Solution to the "Explanatory Gap"

I propose that we need a *two-part solution* to eliminate the "explanatory gap," and both parts entail the issues of emergence and NBE that I discuss in this book. First, it is clear, and no neuroscientist disputes, that the feeling of say the color red versus the pain of a pinprick is the result of a specific neurobiological functions of an individual's nervous system. And I have argued that sentience ("feelings") is the result of *emergent neural functions.* The different feelings that emerge from these brain processes are numerous and enormously complex, and to my mind there is no evidence that they are not entirely neurobiologically natural and specific to their underlying neurobiological substrates.

For instance, as described in the earlier chapters, it is clear that the neural pathways of color processing, sound processing, pain processing, affect, and so on, show enormous neurobiological differences. *Therefore, these diverse sensations should not—and indeed could not—all have the same subjective "feels."*

So I offer that the "why" question as in "why" do neural states "feel" the way that they do has two primary explanations. First, as explained above, this way of framing this problem completely ignores the various diverse "feelings" that correspond to their diverse underlying neurobiological substrates. The first part of answering why red feels "red" versus the sound of a trumpet or a painful pinprick is clear: *They are associated with different neurobiological substrates.*

But there is a second piece of this puzzle, one that critically involves the personal emergence of sentience. That being that we cannot *dissociate* the subjective nature of those states and the "character" of those states from *both* the *emergent neural substrate of these states* described above as well as the *personally living, and emergent, embodied system nature* of those states that I discussed in postulate 1. So I concluded that sentience and "what it's like to be in a neural state" entail these *two related aspects of the emergence of sentience.*

In other words, sentience ("feeling") *as well as* the personal nature and *character* of sentience and "feeling" are *both* emergent features of neurobiologically evolved complex brains, and we can roughly trace in time the evolutionary emergence of sentience from sensing (figure 4.1).

But note that in this model, this coemergence doesn't suddenly and mysteriously appear. It took billions of years to evolve from life to sentience with the emergence of sentience occurring at a minimum about 520–540 million years ago.

An "Experiential Gap"

Thus, I propose that Broad, and Levine, and Chalmers, and the "Mary problem" are really not describing a scientific explanatory gap but rather an *experiential gap*[2] whose two elements are both scientifically explained by the emergent nature of sentience. And therefore, ultimately, attempting to fully "objectify" the subjective *experiential aspects* of sentience is futile.

Conclusions

In conclusion, I propose that we can reliably identify three parallel progressive stages in the emergence of sentience: the increase in *complex neurobiological features*, the extended *evolutionary timeline*, and the eventual emergence of *subjective experience*. I argue that NBE offers a fully naturalized theory of sentience that unifies the biology, neurobiology, evolution, and some aspects of the philosophy of sentience and consciousness without *denying* the personal subjective nature of sentience or trying to fully *objectify* it. Further, this emergentist position neither supports dualism nor is it an argument against physicalism and helps resolves some long-standing debates regarding the nature of consciousness and sentience.

Epilogue: Emergence, Process, and Sentience

There are two meanings of the term "emergent." The first has mystical overtones. It implies that the emergent behavior cannot in any way, even in principle, be understood as the combined behavior of its separate parts. I find it difficult to relate to this type of thinking. The scientific meaning of emergent, or at least the one I use, assumes that, while the whole may not be the simple sum of the separate parts, its behavior can, at least in principle, be understood from the nature and behavior of its parts plus the knowledge of how all these parts interact.

There is no obvious reason why we should not be able to obtain this knowledge—both of the components of the brain and also how they interact together. It is the sheer variety and complexity of the process involved that makes our progress so slow.

—Both quotes are from Francis Crick, 1994, pp. 11–12.

Francis Crick, James Watson, and Maurice Wilkins shared the 1962 Nobel Prize in Physiology or Medicine for their discovery of the molecular structure of the DNA molecule. In his later years, Crick became interested in the problem of the biology of consciousness. The above quotes are from the Introduction to his book *The Astonishing Hypothesis. The scientific search for the soul* that was published in 1994, at a time when scientific studies of consciousness were gaining steam.

He made these observations about emergence in the context of theories about consciousness that are the subject of his book. And I find remarkable agreement with the views that he expressed almost 30 years ago and the postulates that I propose here for Neurobiological Emergentism (NBE).

First, he rejects outright the view that emergence cannot "even in principle" explain consciousness. Second, he endorses a view that consciousness can be explained by the nature and behavior of the "parts" of the brain and the knowledge of how these parts interact. But third, he foresaw that one obstacle to a full emergent explanation would be the "sheer variety and complexity of the processes involved" in the emergence of sentience.

This last observation is very important for my theory of NBE for two reasons. First, as I have described and emphasized in this book, we indeed find that there is enormous diversity and complexity of emergent mechanisms that contribute to consciousness or, more specifically, sentience. And second, he is somewhat unique in that he also notes that we are dealing with not just neural *structures* but the variety and complexity of the *processes* that these structures create.

On this issue, Crick's observation resonates with what I described in chapter 7 regarding what other scientific giants such as William James and Ernst Mayr also noted regarding the importance of process to the emergence of consciousness: that when we consider our consciousness or sentience to be ultimately "only" the *physical* neurons of the brain, we should not forget that sentience is a *process* not a structure.

This does not mean that the physical substrate is not important. Indeed, as I have argued, the actual physical properties of life and the brain are key for an explanation of sentience. What I mean to say is that in order to integrate the structure of the brain with experience, it is essential to understand that emergent processes are just that—*processes*. And that to bring about an emergent aggregate system feature such as sentience, the creation of which spans all manner of "levels" and varieties of organization, we need to consider the emergent aggregate *processes* that nervous systems make possible. In other words, sentience is created by as much what our brains *do* as by what they *are*. So while we rightfully think a brain as a material "thing" when we think of consciousness or sentience, we should think about these as *embodied and personal emergent processes* that are created by the "material" brain.

Process and the Personal Nature of Sentience

So I propose that NBE solves many of the problems of explaining consciousness that Crick pointed out. But another problem that NBE solves that Crick did not address is the personal nature of consciousness and especially sentience.

The idea that sentience has a personal nature is justifiably mysterious. Why can't the personal nature of red or pain or happiness be simply reduced to the material brain? But the answer is surprisingly simple. Sentience is an emergent feature of the life of the organism. And just as a life is irreducibly tied to the physical organism that creates it, the same is true of that organism's "feelings" to whatever degree they exist. So just as life is by its nature an *embodied personal process* of a living organism, so sentience as an emergent feature of the life of the organism and the emergent biological and especially neurobiological processes of its brain. Indeed, *it could not be any other way than uniquely personal to the individual organism.*

The important point is that *the emergent nature of sentience also explains its personal nature.* The neurobiological emergentist approach explains both the hidden complexities of how sentience is created and evolved and at the same time eliminates the "explanatory gap" between the brain and sentience.

Finally, this emergentist view reshapes in some respects how we might view our own sentience. Our sentience—and our "selves"—while subjectively experienced as a stable objective entity is really an illusion. Our sentience is revealed to be in constant flux, a process that is constantly unfolding, always emerging, and never truly the same from one moment to the next.

Notes

Chapter 1

1. Nagel 1974, p. 436.

2. For *sensory consciousness* and *primary consciousness*, see Feinberg and Mallatt 2013, 2016a, 2016b; for *phenomenal consciousness*, see Feinberg and Mallatt 2020; also see Revonsuo 2006, 2010.

Other related terms of interest are *subjectivity*: Metzinger 2003; Nagel 1986; Searle 1992, 1997; Tye 2000; Velmans 2009; or the experience of *qualia*: Chalmers 1995, 1996, 2010; Churchland and Churchland 1981; Dennett 1988; Flanagan 1992; Jackson 1982; Levine 1983; Metzinger 2003; or simply *experience*: Chalmers 1995; Koch 2019.

Philosopher David Chalmers writes that he uses "experience" "conscious experience" and "subjective experience" more or less interchangeably as synonyms for "phenomenal consciousness." Based upon Nagel, he says that "a system is phenomenally conscious if there is something it is like to be that system, from the first-person point of view" (Chalmers 2018, p. 6).

3. https://www.merriam-webster.com/dictionary/sentient

4. Mallatt and Feinberg 2020. A further reason to refer to sentience rather than some other alternative terms is that is often the preferred term in the animal consciousness literature for affective feelings and perceived sensations. For instance, a journal devoted to the question of animal consciousness is named *Animal Sentience: An Interdisciplinary Journal on Animal Feeling*. It's founding editor and editor-in-chief is Stevan Harnad who chose that name. Some of his arguments for the term "sentience" to describe animal consciousness can be found in Harnad 2016; 2021. For a collection of essays on the topic of animal sentience, especially from the perspective of animal rights, see D'Silva and Turner 2006.

5. Carruthers and Schier 2017; Chalmers 1995, p. 207, said the same thing when he argued that in order to explain consciousness "we need and *extra ingredient* in the explanation."

6. Feinberg 2013; Feinberg and Mallatt 2016, 2018a, 2018b, 2019, 2020. The theory of NN identified four apparent gaps between subjective experience and the brain called

"neuroontologically subjective features of consciousness (NSFCs)." These are *referral, mental unity, mental causation,* and *qualia.* Referral means that sensory experiences are perceived, never as if in the brain where they are constructed, but as if in the outside world or inside the body. Mental unity is the gap between the divisible, discontinuous brain that consists of individual neurons and the relatively unified, continuous field of consciousness. Mental causation is the puzzle of how the subjective, seemingly immaterial, and objectively unobservable mind can cause physical effects in the material world; and qualia are the "something it is like to be" experiences of sentience that are the primary subject of this book. For commentary, see Irwin 2020; Lewis 2020.

7. Searle 1992.

8. Feinberg and Mallatt 2016a, 2016b, 2018a, 2019, 2020.

9. Discussions of the applications of emergence to mental operations including consciousness can be found in, but is no means limited to, Andersen et al. 2000; Atmanspacher 2012, 2015; Bedau 1997, 2008; Bedau and Humphreys 2008; Beckermann et al. 2011; beim Graben 2014; Broad 1925; Chalmers 2006; Clayton 2006; Clayton and Davies 2006; Deacon 2011; Feigl 1958; Feinberg 2001, 2012; Feinberg and Mallatt 2016a, 2018a, 2019, 2020; Gibb et al. 2019; Gillett 2002a, 2002b; Kim 1998, 2006; Lewes 1877; Mallatt and Feinberg 2017; Macdonald and Macdonald 2010; McLaughlin 2008; Scott 1995; Searle 1992; Silberstein 2001, 2006, 2017; Sperry 1990; Thompson 2007; Van Gulick 2001; Vision 2017.

10. Dualism with reference to the mind-body takes many forms, but the central idea is that mind and body differ in some fundamentally "physical" way. For an excellent review of all the variations on this topic, see Robinson 2020; Avramides 2020; also see Chalmers 2017, where he offers a theory that argues for a form of dualism that he called "Naturalistic Dualism."

11. Like dualism, there are many philosophical varieties of "physicalism:" but the major point is that unlike dualism, the mind and body are both fundamentally "physical." For reviews, see https://en.wikipedia.org/wiki/Physicalism; Stoljar.

12. Elly Vintiadis (2013b) provides an excellent philosophical discussion of the possibility of the emergence of mind from the brain and defends the view that an emergentist view of conscious is entirely consistent with *naturalism* of the type advocated by Searle.

Chapter 2

1. Ernst Mayr (1982, 2004) was a strong advocate for the role of emergence in creating novel biological features.

2. For discussion of the history of emergence theory in general and Lewes, see Davies 2006 and Clayton 2006.

3. Some books that contain contributions to emergence theory run the gamut across multiple domains. For some collections of these, see *The Re-Emergence of Emergence* by Clayton and Davies 2006; *Emergence: Contemporary Readings in Philosophy and Science* by Bedau and Humphreys 2008; and *The Routledge Handbook of Emergence* by Gibbs et al. 2019. For more biologically oriented approaches, see *Evolving Hierarchical Systems* by Salthe 1985 and *Hierarchy. Perspectives for Ecological Complexity* by Allen and Starr 1982. A general and accessible book with a lot of material on emergence in biological systems see *The Emergence of Everything* by Morowitz 2002.

4. Ahl and Allen 1996; Allen and Starr 1982; Beckermann et al. 2011. Bedau 1997; Bedau and Humphreys 2008; Clayton and Davies 2006; Pattee 1970; Mayr 2004; Salthe 1985, 2008; Vintiadis 2013a.

5. Unity is a feature of numerous theories of the basis of consciousness. For discussions and controversies regarding the unity of consciousness, see Bayne and Chalmers 2003; Bayne 2010; Brook and Raymont 2010; Ellia et al. 2021; Feinberg 2012; Feinberg and Mallatt 2016a, 2018b; Tononi et al. 2016.

In the philosophy of consciousness, Sellars (1963) called this the "grain argument" and pointed out that while subjective experience is homogeneous, unified, and "without grain" (Teller 1992) the physical brain is said to be an objectively "gappy," heterogeneous, discontinuous "conglomerate of spatially discrete events."

6. For the importance of hierarchies in the creation emergent properties, see Ahl and Allen 1996; Allen and Starr 1982; Atmanspacher 2012, 2015; Atmanspacher and beim Graben 2009; Beckermann et al. 2011; Bedau 1997, 2008; Bedau and Humphreys 2008; Bishop and Atmanspacher 2006; beim Graben 2014; Clayton and Davies 2006; Feinberg 2012; Mayr 2004; Pattee 1970; Salthe, 1985.

7. In chapter 4, I discuss different ways that hierarchical "levels" may be defined including hierarchical spatial scale and nested versus nonnested hierarchies in the creation of the emergent neurobiological properties that contribute to sentience.

8. A discussion of the progressive "encephalization" of some sentient features such as *sensory mental images* can be found in Feinberg and Mallatt 2016a, 2016b, 2018a, 2018b.

Chapter 3

1. For more extended discussions of the "other minds" problem, see Avramides 2000, 2020; Godfrey-Smith 2016; Nagel 1995.

2. Feinberg and Mallatt 2016a, 2016b, 2018a, 2019, 2020.

3. In the philosophical literature, all of these feelings are sometimes generally referred to as "qualia." See philosopher Michael Tye (2021) for an excellent summary of the meaning of the term "qualia" and its philosophical implications.

4. Edelman 1992, p. 112.

5. Damasio 2000, 2010.

6. For an extended review on the neuroscience of affect, see Adolph and Anderson 2018. For some theories that discuss or focus on affective or interoceptive awareness, see Cabanac 1996; Cabanac et al. 2009; Damasio 2000, 2010; Denton 2006; Panksepp 1998, 2005.

7. See also Anderson and Adolphs 2014; Berridge 2018; Berridge and Kringelbach 2015; Damasio and Carvalho 2013.

8. Lyon 2015.

9. Birch et al. 2021; Crump 2022a, 2022b. See also Birch 2017; Birch et al. 2020, 2021, 2022. For outside scientific commentary on these criteria from a range of perspectives as well as the author's responses, see Crump et al. 2022a.

10. Feinberg and Mallatt 2016a; Mallatt and Feinberg 2022.

11. See also Crook and Walters 2011; Sneddon 2019.

12. For the neurobiology and features of reflexes, see Fischer and Truog 2015; Hultborn 2006; Kandel et al. 2021; Price 2005; Reflexes https://en.wikipedia.org/wiki /Reflex; Definition of REFLEX" *www.merriam-webster.com.*

13. For more on these distinctions, see Crump et al. 2022a.

14. For some general references on types of learning and conditioning in general, see https://en.wikipedia.org/wiki/Learning#Associative_learning; https://www.britannica .com/science/animal-learning/Types-of-learning#ref320590

As we have suggested elsewhere (Ginsburg and Jablonka 2019), open-ended associative learning (in this case of composite predictors of aversive or aversion-ameliorating stimuli or actions)—which requires multimodal discrimination, motivational tradeoffs, instrumental goal-directed conditioning, trace conditioning, and second-order learning—is a very strong indicator of sentience, accompanied invariably by integrative brain areas that support it. Such learning has been shown to be possible (in humans) only when there is conscious awareness (as noted also in the target article). Online updating and prioritizing requires the same functional cognitive-affective architecture.

15. Jablonka and Ginsburg 2022, pp. 2–3; Ginsburg and Jablonka 2019. For other evolutionary analysis of sentience and learning, see Feinberg and Mallatt 2016a; For associative learning and its relationship to consciousness in humans, see Skora et al. 2021.

Chapter 4

1. Some valuable analyses of the relationship between complexity and emergence across different types of complex systems can be found in Bechtel and Richardson

2011; Bishop and Silberstein 2019; Goldstein 2018; Ladyman and Wiesner 2020; Mitchell 2009; Nunez 2016; Simon 1962, 1973; Tononi and Edelman 1998.

2. For emergence and self organization, see Camazine et al. 2001

Chapter 5

1. Mayr 1982, 2004; Salthe 1985; Morowitz 2002;; Van Kranendonk et al. 2017.

2. https://en.wikipedia.org/wiki/Protocell; Chen and Walde 2010; Schrum et al. 2010.

3. Cooper 2000; Hallmann and French 2015; https://sunyorange.edu/biology/resour ces/library/prehistoriclife/prokaryotes.html

4. https://en.m.wikipedia.org/wiki/Chemotaxis; Wadhams and Armitage 2004.

5. Bi and Sourjik 2018; Booth 2014; Haswell et al. 2011; Keymer et al. 2006; Micali and Endres 2016; Partridge et al. 2019; Sourjik and Wingreen 2012; Sagawa et al. 2014; Wadhams and Armitage 2004.

6. According to Wadhams and Armitage (2004), the most common mechanisms entail a complex *signal transduction systems* that consists of a *sensory mechanism* that detects a chemostimulus on the bacterial outer membrane and a regulator mechanism that mediates the bacterium's responses. They note, however, that there are many variations on these basic mechanisms.

7. Kearns 2010; Macnab 1999; Patteson et al. 2015; Sagawa et al. 2014; Sourjik and Wingreen 2012.

8. Reber 2019.

9. Dexter et al. 2020; Gunawardena 2022; Kumazawa 2002; Slabodnick and Marshall 2014; Tartar 1961; Trinh et al. 2019.

10. Bell et al. 1998; Dussutour 2021; Levandowsky et al. 1984; Oami 1996; Snyder 1991; Song et al. 1991; Van Houten et al. 2000.

11. Jennings 1906.

12. Reynierse and Walsh 1967.

13. Dexter et al. 2020; Trinh et al. 2019. For popular accounts of this, see Frazer 2021; Williams 2019.

14. Dexter et al. 2020, p. 4324; For the heterogeneity of the hierarchy, see also Trinh et al. 2019.

15. Woodruff 2016.

Chapter 6

1. Müller et al. 2007; Zumberge et al. 2018.

2. For more on the phylogeny of jellyfish, see Cartwright et al. 2007; Collins 2009; Peterson and Butterfield 2005.

3. Bosch et al. 2017; Katsuki and Greenspan 2013.

4. Evans et al. 2020.

5. Arendt et al. 2016, 2019.

6. For more detail on this evolutionary scenario, see Feinberg and Mallatt 2016a, 2018a, 2018b, 2020.

7. Schafer (2016) provides an excellent and accessible introduction and summary of the general and neurobiological features of *C. elegans*.

8. Ardiel and Rankin 2010; Sulston 1983; Sulston and Horvitz 1977; White et al. 1986.

9. Schafer 2016.

10. Bumbarger et al. 2009; Dekkers et al. 2021; Iliff and Xu 2020; Nishijima and Maruyama 2017; Schafer 2016.

11. Mills et al. 2016; Summers et al. 2015; Tracey 2017; Witham et al. 2016; Wittenburg and Baumeister 1999;

12. Brittin et al. 2021.

13. Metaxakis et al. 2018.

14. Fang-Yen et al. (2015) and Ghosh et al. (2017) come to similar conclusions.

15. Butler and Hodos 2005; Glover Fritzsch 2009; Nieuwenhuys 1998.

16. Benito-Gutiérrez et al. 2021; Holland 2015; Holland and Chen 2001; Lacalli, 1996, 2004, 2008, 2010, 2021, 2022; Stokes 1997; Stokes and Holland 1995; Wicht and Lacalli 2005;

For adult neuron numbers, see Candiani et al., 2012; Nicol and Meinertzhagen 1991. The estimate of the number of neurons in the larval stage is from Lacalli (personal communication).

17. Churcher and Taylor 2009; Poncelet and Shimeld 2020.

18. Pergner and Kozmik 2017; Lacalli 2018a, 2018b.

19. Lacalli 2017, 2018a, 2018b, 2021; Lacalli & Candiani 2017.

20. Ponder and Lindberg 2008.

21. Moroz 2011.

22. For the anatomy of gastropod mollusks, see https://case.edu/artsci/biology/chiellab /about/our-research; Moroz 2011; Moroz et al. 2006, 2014; https://en.wikipedia.org

/wiki/Molluska; https://case.edu/artsci/biology/chiellab/about/our-research; https://aplysia.rsmas.miami.edu/biology-and-production-of-aplysia%20/biological-description/index.html

23. Gillette and Brown 2015; Illich and Walters 1997; Wertz et al. 2006; Wollesen et al. 2007.

24. Gillette and Brown 2015; see also Catanho et al. 2012; Hirayama and Gillette 2012; Hirayama et al. 2012, 2014; Jing and Gillette 2000; Noboa and Gillette 2013.

25. Mills et al. 2016.

26. Talkington 2001.

27. Feinberg and Mallatt 2016a.

28. Holland and Yu 2002.

29. Feinberg and Mallatt 2016a.

30. Benito-Gutiérrez et al. 2021

31. See also Gillette and Brown 2015.

Chapter 7

1. Indeed, the lack of extensive cross-communication is considered by some theories to account for why some brain regions such as the cerebellum can coordinate complex motor movements but operate largely nonconsciously. Koch 2019; Koch et al. 2016; Tononi and Koch 2015.

2. Adolphs and Anderson 2018; Feinberg and Mallatt 2016a, 2018a.

3. See also Mayr 2004.

4. James 1904. For an historical perspective and analysis of James's work, see the Macat Team 2017.

5. Feinberg and Mallatt 2013, 2016a, 2016b, 2018a; Mallatt and Feinberg 2021.

6. For more detailed analysis of the evolution of exterosensory mental images, see Feinberg and Mallatt 2013, 2016a, 2016b, 2018a.

7. Gans and Northcutt 1983; Hall 2008; Northcutt 2005; Schlosser et al. 2014.

8. Feinberg and Mallatt 2016a. For the mesolimbic reward system and the neuroanatomical aspects of valence, see Adolph and Anderson 2018; Berridge and Kringelbach 2015; Feinstein 2013; Goodson and Kingsbury 2013; Ikemoto 2010; Kender et al. 2008; Kringelbach and Berridge 2017; O'Connell and Hofmann 2012.

9. Feinberg and Mallatt 2016a, 2018a.

10. Bulbert et al. 2015.

11. Birch et al. 2020, 2021; Carls-Diamante 2017; Godfrey-Smith 2016; Mather 2021a, 2021b; Montgomery 2015; Ponte et al. 2022.

12. Edelman and Seth 2009; Hanlon and Messenger 2018; Hochner 2008, 2012; Hochner et al., 2006; Shigeno and Ragsdale 2015; Shigeno et al. 2018; Young 1971, 1974; Zullo and Hochner 2011; Zullo et al. 2009.

13. Edelman and Seth 2009; For senses of cephalopods in general and/or octopus, specifically see Visual: Mather 2021b; Shigeno and Ragsdale 2015; Williamson and Chrachri 2004; Young 1974; Mechanosensory: Grasso 2014; Mather 2012; Shigeno and Ragsdale 2015; Olfaction: Polese et al. 2015, 2016; Auditory: Kaifu et al. 2007; Mather 2012; Taste: Grasso 2014; Graziadei 1964; Mather 2012; Wells 1963.

14. Pungor et al. 2023.

15. Chung et al. 2020.

16. For more on the issue of sensory and motor somatotopy in the octopus, see Carls-Diamante 2022; Hochner 2012; Shigeno et al. 2018; Zullo et al. 2009.

17. See Birch et al. 2021 for review and references for cephalopod nociceptors.

18. Shigeno et al. 2018 refers to it as *frontal-vertical lobe* or more simply the *vertical lobe* (figure 6.2). For their role in cognitive and memory functions, see Edeleman and Seth 2009; Maldonado 1965; Shigeno et al. 2018; Shomrat et al. 2015; Wells and Young 1969; Williamson and Chrachri 2004; Young 1961, 1971, 1972, 1974, 1991, 1995.

19. Shigeno et al. 2018; see also Young 1991, 1995. Shigeno also notes (personal communication) that further support for the mammalian cerebral cortex analogy comes from a similar expansion of areas gyri (folds) and layers (they used the terms zones or lamina as in insect mushroom bodies) in both octopuses and vertebrates (Shigeno and Ragsdale 2015).

20. Crook 2021.

21. Birch et al. 2021; See also Gutnick et al. 2020; Hanlon and Messenger 2018; Hochner et al. 2006; Marini et al. 2017; Mather 1995, 2008, 2012; Schnell et al. 2021.

22. Godfrey-Smith 2016, 2019a, 2019b; Gutnick et al. 2011; Hanlon and Messenger 2018; Hochner 2008, 2021a, 2021b; Mather 2019; Mikhalevich and Powell 2020; Montgomery 2015; Packard and Delafield-Butt 2014; Papini and Bitterman 1991. Edelman and Seth 2009; Ponte et al. 2022.

For a comprehensive review and listing of relevant papers, see Schnell et al. 2021.

23. Schnell et al. 2021. Also for behavioral flexibility, see Mikhalevich and Powell 2020.

24. Messenger 2001.

25. Hanlon et al. 2008; Hanlon et al. 1999, 2010; Huffard 2006; Moynihan and Rodaniche 1982; Norman et al. 1999, 2001; Panetta et al. 2017.

26. Finn et al. 2009.

27. For numbers of neurons for insects versus vertebrates, see Chittka and Niven 2009; http://en.wikipedia.org/wiki/List_of_animals_by number_of_neurons

28. For the general neuroanatomy of the insect brain, see Strausfeld 2009, 2013.

For senses and somatotopic organization in insects, see Vision: Seelig and Jayaraman 2013; Sombke and Harzsch 2015; Olfaction: Jacobson and Friedrich 2013; Mamlouk and Schmuker 2011; Strausfeld 2013; Taste: Newland et al. 2000; Wolf 2008; Auditory: Strausfeld 2013; Mechanosensory: Strausfeld 2013; Newland et al. 2000.

29. For an analysis of insect sensory neurohierarchical pathways in general, see Feinberg and Mallatt 2016a.

30. Seelig and Jayaraman 2013.

31. Eichler et al. 2017; Iwano et al. 2010.

32. Solvi et al. 2020.

33. Boto et al. 2020; Chittka 2022; Eichler et al. 2017; Hattori et al. 2017; Li and Strausfeld 1999; Strausfeld et al. 2020; Takemura et al. 2017.

34. Aso et al. 2014a, 2014b; Cognigni et al. 2018; Eschbach et al. 2021; Owald et al. 2015; Siju et al. 2020; Yamazaki et al. 2018.

35. Chittka 2022; Key et al. 2021; Wolf et al. 2015.

36. Hu et al. 2018.

37. For some good discussions and controversies regarding pain in insects, see Adamo 2016; 2019; Key et al. 2021. Shelly Adamo argues that when compared to the infrastructure of human pain processing, the insect condition shows much less integration cross-communication between the mushroom bodies and central complex. But on this issue, and we agree, she does concede that there are questions as to what degree comparisons between the neural infrastructure of insects and humans can be applied.

38. Gibbons et al. 2022c; see also Gibbons et al. 2022a, 2022b.

39. For operantly learned responses to punishments or rewards, see Brembs 2003a, 2003b, 2014; Brembs and Heisenberg 2000; Heisenberg et al. 2001; Putz and Heisenberg 2002; Wustman and Heisenberg 1997; Wustman et al. 1996; For self-delivery of analgesics or rewards, see: Søvik and Barron 2013; and for approach to reinforcing drugs, see Shohat-Ophir et al. 2012.

40. Bateson et al. 2011.

41. Solvi et al. 2016.

42. Gibbons et al. 2022a, 2022b, 2022c.

43. For some papers regarding cognition, nonreflexive behaviors, and sentience in insects, see Baracchi and Baciadonna 2020; Barron and Klein 2016; Chittka and Niven 2009; Chittka and Wilson 2019; Giurfa 2013; Klein and Barron 2016; Loukola et al. 2017; Solvi et al. 2016, 2020; Swinderen 2005. For the complexity and flexibility of insect escape behaviors that we have discussed earlier in other species, see Card 2012; Card and Dickinson 2008a, 2008b.

44. Perry and Chittka 2019.

45. For which species show which of these behaviors, see the original article (Perry and Chitta 2019). See also Nityananda and Chitka 2021; Overgaard 2021.

46. Loukola et al. 2017.

47. For discussion of the visual system, see Jenkins et al. 2022; for olfactory, see Harzsch and Krieger 2018; Jenkins et al. 2022; Krieger et al. 2015.

48. Krieger et al. 2012.

49. Strausfeld 2020; Strausfeld et al. 2020.

50. Birch et al. 2021; Crump et al. 2022. The Crump et al. 2022 paper reviews the Birch et.al. findings focusing on decapod crustaceans.

51. Fanjul et al. 2008; Spivak 2010.

52. Tomsic et al. 2017; see also Hemmi and Tomsic 2012; Medan et al. 2007; Oliva et al. 2007; Scarano and Tomsic 2014. Similar escape behavior patterns have also been found in fiddler crabs: Hemmi 2005; Smolka et al. 2013. For a more general review of decision making and behavioral choice among other arthropod species, see Herberholz and Marquart 2012.

53. Tomsic and Maldonado 2014.

54. Elwood 2022.

55. Elwood 2022.

56. Chase 1991; Chase and DeWitt 1988; Chase et al. 1988; Laidre 2014, 2021.

57. Mayer et al. 2010; Martin et al. 2022; Strausfeld et al. 2006; Vizueta et al. 2020.

58. Differences in the brain divisions, see Mayer et al. 2010; homologies of mushroom bodies and central bodies, see Martin et al. 2022.

59. Martin et al. 2022; Storch and Ruhberg 1977, 1993.

50. Kirwan et al. 2018; Martin et al. 2022.

51. Reinhard and Rowell 2005.

Chapter 8

1. Sperry 1980, 1984, 1990, 1991. See Feinberg 2001 for an earlier discussion of Sperry's theory.

2. Sperry 1980, 1991.

3. See the Preface for discussions and references on dualism and strong emergence.

4. Chalmers 2017.

5. Revonsuo 2010; For other discussions of mental causation see also Dardis 2008; Feinberg 2012; Heil and Mele 1993; Walter and Heckmann 2003.

6. Chalmers 2006; See also see also Emmeche et al. 2000.

7. Brooks et al. 2021.

8. Atmanspacher 2012, 2015; Atmanspacher and beim Graben, 2009; beim Graben 2014; Craver 2015; Povich and Craver 2017; Silberstein 2001, 2006, 2017.

9. Brooks 2017; Wimsatt 1994/ 2007; Potochnik and McGill 2012.

10. Hilgetag and Goulas (2020) proposed several ways to describe this type of "sequential" neural hierarchy in the visual system of the primate brain. One of these is based upon the *laminar projection patterns* in which the levels of the hierarchy are organized by of ascending ("lower") descending "higher" projections. A second related version is based solely upon the shortest synaptic pathway from the sensory receptor (in the case the retina) and the "higher" brain regions. Both of these sorts of hierarchies most clearly apply to the topographical sensory pathways described above.

11. For more on nested hierarchies, see Allen and Starr 1982; Eronen 2021; Feinberg 2012; Feinberg and Mallatt 2013; Salthe 1985.

12. There is sizable literature on HMN in neural systems. For an excellent review and discussion, see Sporns 2011. Also see Hilgetag and Goulas 2020; Hilgetag and Hütt 2014; Kaiser and Hilgetag 2010; Kaiser et al. 2007, 2010; Meunier et al. 2009.

13. Sporns 2011.

14. See Nunez 2016 for a review of these temporal patterns.

15. Singer is quoted in Aulette et al. 2013.

16. For a compilation on the topic of downward causation, see Andersen et al. 2000.

17. Figure is adapted from Lewin 1992, p. 13.

18. Hansen 2015; Mallatt and Feinberg 2021.

19. Ahl and Allen 1996; Salthe 1985; Simon 1962, 1973.

20. Nunez 2016. Also see Noble et al. 2019.

Chapter 9

1. Haekel 1892; Thompson 2022.

2. Oizumi et al. 2014; Koch and Laurent 1999; Tononi 2015, 2017; Tononi and Koch 2015; Tononi et al. 2016. For a review and commentary on ITT, see Merker et al. 2022.

3. For some earlier useful comparisons between or original theory of NN and ITT, see Mallatt 2021.

4. Baars et al. 2013; Ellia and Chis-Ciure 2022; Feinberg and Mallatt, 2016; Lamme 2006; Min 2010; Mallatt 2021.

5. Doerig et al. 2021; Fallon 2016; Michel et al. 2019.

6. Mallatt 2021.

7. Horgan 2015; Koch 2019; Tononi 2008; Van Gulick 2019, 2021.

Chapter 10

1. Block 1995; Broad, 1925; Carruthers 2016; Choifer, 2018; Chalmers, 1995, 1996; Conee 1994; Hasan and Fumerton, 2019; Jackson 1982, 1986; Levine 1983; Metzinger 1995, 2003; Nagel 1974, 1986; Nida-Rümelin and Conaill 2019; Revonsuo 2006; Russell 1910–1911, 1912, 1914; Searle 1992, 2007; Shear 1999; Teller 2011; Tye 2002; Velmans 1991.

2. Feinberg and Mallatt 2016a, 2016b, 2018a, 2019, 2020.

3. Feinberg and Mallatt 2020.

4. See, for instance, the "knowledge argument" against physicalism: Conee 1994; Jackson 1982, 1986; Nida-Rümelin and Conaill 2019.

5. Jackson 1982, 1986, 2004; Ludlow et al. 2004.

https://plato.stanford.edu/entries/qualia-knowledge

6. For some discussions of *knowledge by acquaintance* see, for instance, Conee 1994; Feinberg and Mallatt 2020; Hasan and Fumerton 2019; Jackson 1982, 1986; Metzinger 2003; Nida-Rümelin and O Conaill 2019; Tye 2002; Searle 1992, 2007.

7. See Ludlow et al. 2004 for a collection of chapters on the Mary problem.

Chapter 11

1. The diversity of the types of nervous systems that can create sentience has been called "multiple realizabilty" (Shapiro 2004; Polger and Shapiro 2016). For a review and analysis of this idea in cross-species comparisons, see Mallatt and Feinberg 2021.

2. Feinberg and Mallat 2020.

References

Adolphs, R., & Anderson, D. J. (2018). *The neuroscience of emotion: A new synthesis.* Princeton University Press.

Adamo, S. A. (2016). Do insects feel pain? A question at the intersection of animal behaviour, philosophy and robotics. *Animal Behaviour, 118,* 75–79.

Adamo, S. A. (2019). Is it pain if it does not hurt? On the unlikelihood of insect pain. *The Canadian Entomologist, 151*(6), 685–695.

Ahl, V., & Allen, T. F. H. (1996). *Hierarchy theory.* Columbia University Press.

Allen, T. F. H., & Starr, T. B. (1982). *Hierarchy: Perspectives for ecological complexity.* University of Chicago Press.

Anderson, D. J., & Adolphs, R. A. (2014). Framework for studying emotions across species. *Cell, 157,* 187–200.

Andersen, P. B., Emmeche, C., Finnemann, N. O., & Christiansen, P. V. (Eds.). (2000). *Downward causation: Minds, bodies and matter.* Aarhus University Press.

Ardiel, E. L., & Rankin, C. H. (2010). An elegant mind: Learning and memory in *Caenorhabditis elegans. Learning & Memory, 17*(4), 191–201.

Arendt, D., Bertucci, P. Y., Achim, K., & Musser, J. M. (2019). Evolution of neuronal types and families. *Current Opinion in Neurobiology, 56,* 144–152.

Arendt, D., Tosches, M. A., & Marlow, H. (2016). From nerve net to nerve ring, nerve cord and brain—Evolution of the nervous system. *Nature Reviews Neuroscience, 17*(1), 61–72.

Aso, Y, Sitaraman, D, Ichinose, T, Kaun, K. R., Vogt, K., Belliart-Guérin, G, Plaçais, P. Y., Robie, A. A., Yamagata, N., Schnaitmann, C., Rowell, W. J., Johnston, R. M., Ngo, T. T., Chen, N., Korff, W., Nitabach, M. N., Heberlein, U., Preat, T., Branson, K. M., Tanimoto, H., et al. (2014a). Mushroom body output neurons encode Valence and guide memorybased action selection in Drosophila. *eLife, 3,* e04580.

Aso, Y., Hattori, D., Yu, Y., Johnston, R. M., Iyer, N. A., Ngo, T. T., Dionne, H., Abbott, L. F., Axel, R., Tanimoto, H., & Rubin, G. M. (2014b). The neuronal architecture of the mushroom body provides a logic for associative learning. *eLife, 3*, e04577.

Atmanspacher, H. (2012). Identifying mental states from neural states under mental constraints. *Interface Focus, 2*, 74–81.

Atmanspacher, H. (2015). Contextual emergence of mental states. *Cognitive Processing, 16*(4), 359–364.

Atmanspacher, H., & beim Graben, P. (2009). Contextual emergence. *Scholarpedia, 4*(3), 7997.

Auletta, C., & Jeannerod, M. (Eds.). (2013). *Brains top down: Is top-down causation challenging neuroscience?* World Scientific.

Avramides, A. (2000). *Other minds*. Routledge.

Avramides, A. (2020). Other minds. *The Stanford encyclopedia of philosophy* (Winter 2020 Edition), Edward N. Zalta (Ed.), https://plato.stanford.edu/archives/win2020 /entries/other-minds/

Baars, B. J., Franklin, S., & Ramsøy, T. Z. (2013). Global workspace dynamics: Cortical "binding and propagation" enables conscious contents. *Frontiers in Psychology, 4*, 200.

Baracchi, D., & Baciadonna, L. (2020). Insect sentience and the rise of a new inclusive ethics. *Animal Sentience 5*(29), 18.

Bargmann, C. I. (2006). Chemosensation in *C. elegans*. In *WormBook: The Online Review of C. elegans Biology [Internet]*. WormBook; 2005–2018.

Barron, A. B., & Klein, C. (2016). What insects can tell us about the origins of consciousness. *Proceedings of the National Academy of Sciences, 113*(18), 4900–4908.

Bateson, M., Desire, S., Gartside, S. E., & Wright, G. A. (2011). Agitated honeybees exhibit pessimistic cognitive biases. *Current Biology, 21*(12), 1070–1073.

Bayne T. (2010). *The unity of consciousness*. Oxford University Press.

Bayne, T., & Chalmers, D. J. (2003). What is the unity of consciousness? In A. Cleeremans (Ed.), *The unity of consciousness: Binding, integration, and dissociation* (pp. 23–58). Oxford University Press.

Bechtel, W., & Richardson, R. C. (2011). Emergent phenomena and complex systems. In A. Beckermann, H. Flohr, & J. Kim (Eds.), *Emergence or reduction? Essays on the prospects of nonreductive physicalism* (pp. 257–288). Walter de Gruyter.

Beckermann, A., Flohr, H., & Kim, J. (Eds.). (2011). *Emergence or reduction? Essays on the prospects of nonreductive physicalism*. Walter de Gruyter.

Bedau, M. A. (1997). Weak emergence. *Philosophical Perspectives, 11*, 375–399.

Bedau, M. A. (2008). Downward causation and the autonomy of weak emergence. In M. A. Bedau & P. Humphreys (Eds.), *Emergence: Contemporary readings in philosophy and science* (pp. 155–188). MIT Press.

Bedau, M. A., & Humphreys, P. (2008). *Emergence: Contemporary readings in philosophy and science.* MIT Press.

beim Graben, P. (2014). Contextual emergence of intentionality. *Journal of Consciousness Studies, 21*(5–6), 75–96.

Bell, W. E., Karstens, W., Sun, Y., & Van Houten, J. L. (1998). Biotin chemoresponse in Paramecium. *Journal of Comparative Physiology A, 183*(3), 361–366.

Benito-Gutiérrez, È., Gattoni, G., Stemmer, M., Rohr, S. D., Schuhmacher, L. N., Tang, J., . . . & Arendt, D. (2021). The dorsoanterior brain of adult amphioxus shares similarities in expression profile and neuronal composition with the vertebrate telencephalon. *BMC Biology, 19*(1), 1–19.

Berridge, K. C. (2018). Evolving concepts of emotion and motivation. *Frontiers in Psychology, 9*, 1647.

Berridge, K. C. (2019). Affective valence in the brain: Modules or modes? *Nature Reviews Neuroscience, 20*(4), 225–234.

Berridge, K. C., & Kringelbach, M. L. (2015). Pleasure systems in the brain. *Neuron, 86*(3), 646–664.

Bi, S., & Sourjik, V. (2018). Stimulus sensing and signal processing in bacterial chemotaxis. *Current Opinion in Microbiology, 45*, 22–29.

Birch, J. (2017). Animal sentience and the precautionary principle. *Animal Sentience, 2*(16), 1.

Birch, J., Broom, D. M., Browning, H., Crump, A., Ginsburg, S., Halina, M., . . . & Zacks, O. (2022). How should we study animal consciousness scientifically? *Journal of Consciousness Studies, 29*(3–4).

Birch, J., Burn, C., Schnell, A., Browning, H., & Crump, A. (2021). *Review of the evidence of sentience in cephalopod molluscs and decapod crustaceans.* London School of Economics and Political Science.

Birch, J., Schnell, A. K., & Clayton, N. S. (2020). Dimensions of animal consciousness. *Trends in Cognitive Sciences, 24*(10), 789–801.

Bishop, R., & Silberstein, M. (2019). Complexity and feedback. In S. Gibb, R. F. Hendry, & T. Lancaster (eds.), *The Routledge handbook of emergence* (pp. 145–156). Routledge.

Bishop, R. C., & Atmanspacher, H. (2006). Contextual emergence in the description of properties. *Foundations of Physics, 36*(12), 1753–1777.

Block, N. (1995). On a confusion about a function of consciousness. *Behavioral and Brain Sciences*. *18*(2), 227–247.

Booth, I. R. (2014). Bacterial mechanosensitive channels: Progress towards an understanding of their roles in cell physiology. *Current Opinion in Microbiology, 18*, 16–22.

Bosch, T. C., Klimovich, A., Domazet-Lošo, T., Gründer, S., Holstein, T. W., Jékely, G., et al. (2017). Back to the basics: Cnidarians start to fire. *Trends in Neuroscience, 40*, 92–105.

Boto, T., Stahl, A., & Tomchik, S. M. (2020). Cellular and circuit mechanisms of olfactory associative learning in Drosophila. *Journal of Neurogenetics, 34*(1), 36–46.

Brembs, B. (2003a). Operant conditioning in invertebrates. *Current Opinion in Neurobiology, 13*(6), 710–717.

Brembs, B. (2003b). Operant reward learning in Aplysia. *Current Directions in Psychological Science, 12*(6), 218–221.

Brembs, B. (2014). Aplysia operant conditioning. *Scholarpedia, 9*(1), 4097.

Brembs, B., & Heisenberg, M. (2000). The operant and the classical in conditioned orientation of Drosophila melanogaster at the flight simulator. *Learning & Memory, 7*(2), 104–115.

Brittin, C. A., Cook, S. J., Hall, D. H., Emmons, S. W., & Cohen, N. (2021). A multiscale brain map derived from whole-brain volumetric reconstructions. *Nature, 591*(7848), 105–110.

Broad, C. D. (1925). *The mind and its place in nature*. Routledge.

Brooks, D. S. (2017). In defense of levels: Layer cakes and guilt by association. *Biological Theory, 12*(3), 142–156.

Brooks, D. S., DiFrisco, J., & Wimsatt, W. C. (Eds.). (2021). *Levels of organization in the biological sciences*. MIT Press.

Brook, A., & Raymont, P. (2010). The unity of consciousness. In E. N. Zalta (ed.), *The Stanford encyclopedia of philosophy*. http://plato.stanford.edu/archives/fall2010/entries/consciousness-unity

Bulbert, M. W., Page, R. A., & Bernal, X. E. (2015). Danger comes from all fronts: Predator-dependent escape tactics of Túngara frogs. *PLoS One, 10*(4), e0120546.

Bumbarger, D. J., Wijeratne, S., Carter, C., Crum, J., Ellisman, M. H., & Baldwin, J. G. (2009). Three-dimensional reconstruction of the amphid sensilla in the microbial feeding nematode, *Acrobeles complexus* (Nematoda: Rhabditida). *Journal of Comparative Neurology, 512*(2), 271–281.

Butler, A. B. (2008). Evolution of brains, cognition, and consciousness. *Brain Research Bulletin, 75*(2), 442–449.

Butler, A. B., & Cotterill, R. M. (2006). Mammalian and avian neuroanatomy and the question of consciousness in birds. *Biological Bulletin, 211*(2), 106–127.

Butler, A. B., & Hodos, W. (2005). *Comparative vertebrate neuroanatomy* (2nd ed.). Wiley Interscience.

Cabanac, M. (1996). On the origin of consciousness, a postulate and its corollary. *Neuroscience and Biobehavioral Reviews, 20*, 33–40.

Cabanac, M., Cabanac, A. J., & Parent, A. (2009). The emergence of consciousness in phylogeny. *Behavioural Brain Research, 198*(2), 267–272.

Camazine, S., Deneubourg, J., & Franks, N. R., et al. (2001). *Self-organization in biological systems*. Princeton University Press.

Candiani, S., Moronti, L., Ramoino, P., Schubert, M., & Pestarino, M. (2012). A neurochemical map of the developing amphioxus nervous system. *Bmc Neuroscience, 13*(1), 1–17.

Card, G. M. (2012). Escape behaviors in insects. *Current Opinion in Neurobiology, 22*(2), 180–186.

Card, G., & Dickinson, M. (2008a). Performance trade-offs in the flight initiation of Drosophila. *Journal of Experimental Biology, 211*(3), 341–353.

Card, G., & Dickinson, M. H. (2008b). Visually mediated motor planning in the escape response of Drosophila. *Current Biology, 18*(17), 1300–1307.

Carls-Diamante, S. (2017). The octopus and the unity of consciousness. *Biology & Philosophy, 32*(6), 1269–1287.

Carls-Diamante, S. (2022). Where is it like to be an octopus? *Frontiers in Systems Neuroscience, 16*, 840022.

Carruthers, G., & Schier, E. (2017). Why are we still being hornswoggled? Dissolving the hard problem of consciousness. *Topoi, 36*(1), 67–79.

Carruthers, P. (2016). Higher-order theories of consciousness. *The Stanford encyclopedia of philosophy.* https://plato.stanford.edu/archives/fall2016/entries/consciousness-higher/

Cartwright, P., Halgedahl, S. L., Hendricks, J. R., Jarrard, R. D., Marques, A. C., Collins, A. G., & Lieberman, B. S. (2007). Exceptionally preserved jellyfishes from the Middle Cambrian. *PloS One, 2*(10), e1121.

Chalmers, D. J. (1995). Facing up to the problem of consciousness. *Journal of Consciousness Studies, 2*(3), 200–219.

Chalmers, D. J. (1996). *The conscious mind: In search of a fundamental theory.* Oxford University Press.

Chalmers, D. J. (2006). Strong and weak emergence. In P. Clayton & P. Davies (Eds.), *The re-emergence of emergence* (pp. 244–254). Oxford University Press.

Chalmers, D. J. (2010). *The character of consciousness*. Oxford University Press.

Chalmers, D. (2017). Naturalisic dualism. In M. Velmans & M. Schneider (Eds.), *The Blackwell companion to consciousness* (pp. 359–368). Wiley–Blackwell.

Chalmers, D. (2018). The meta-problem of consciousness. *Journal of Consciousness Studies, 25*(9–10), 6–61.

Chase, I. D. (1991). Vacancy chains. *Annual Review of Sociology, 17*(1), 133–154.

Chase, I. D., & DeWitt, T. H. (1988). Vacancy chains: A process of mobility to new resources in humans and other animals. *Social Science Information, 27*(1), 83–98.

Chase, I. D., Weissburg, M., & Dewitt, T. H. (1988). The vacancy chain process: A new mechanism of resource distribution in animals with application to hermit crabs. *Animal Behaviour, 36*(5), 1265–1274.

Chen, I. A., & Walde, P. (2010). From self-assembled vesicles to protocells. *Cold Spring Harbor Perspectives in Biology, 2*(7), a002170.

Chittka, L. (2022). *The mind of a bee*. Princeton University Press.

Chittka, L., & Niven, J. (2009). Are bigger brains better? *Current Biology, 19*(21), R995–R1008.

Chittka, L., & Wilson, C. (2019). Expanding consciousness. *American Scientist, 107*, 364–369.

Choifer, A. (2018). A new understanding of the first-person and third-person perspectives. *Philosophical Papers* 47, 333–371. doi: 10.1080/05568641.2018.1450160Churc her, A. M., & Taylor, J. S. (2009). Amphioxus (*Branchiostoma floridae*) has orthologs of vertebrate odorant receptors. *BMC Evolutionary Biology, 9*(1), 1–10.

Chung, W. S., Kurniawan, N. D., & Marshall, N. J. (2020). Toward an MRI-based mesoscale connectome of the squid brain. *Iscience, 23*(1).

Clayton, P. (2006). Conceptual foundations of emergence theory. In P. Clayton & P. Davies (eds.), *The re-emergence of emergence* (pp. 1–31). Oxford University Press.

Clayton, P., & Davies, P. (2006). *The re-emergence of emergence: The emergentist hypothesis from science to religion* (No. 159). Oxford University Press.

Cognigni, P., Felsenberg, J., & Waddell, S. (2018). Do the right thing: Neural network mechanisms of memory formation, expression and update in Drosophila. *Current Opinion in Neurobiology, 49*, 51–58.

Collins, A. G. (2009). Recent insights into cnidarian phylogeny. *Smithsonian Contributions to the Marine Sciences, 38*, 139–149.

Conee, E. (1994). Phenomenal knowledge. *Australasian Journal of Philosophy, 72,* 136–150.

Cooper G. M. (2000). The origin and evolution of cells. In *The cell: A molecular approach,* 2nd ed. Sinauer Associates. https://www.ncbi.nlm.nih.gov/books /NBK9839/

Courtiol, E., & Wilson, D. A. (2014). Thalamic olfaction: Characterizing odor processing in the mediodorsal thalamus of the rat. *Journal of Neurophysiology, 111*(6), 1274–1285.

Craver, C. F. (2015). Mechanisms and emergence. In *Open MIND.* MIND Group.

Crick, F. H. C. (1994). *The astonishing hypothesis: The scientific search for the soul.* Charles Scribner's Sons.

Crook, R. J. (2021). Behavioral and neurophysiological evidence suggests affective pain experience in octopus. *Iscience, 24*(3), 102229.

Crook, R. J., & Walters, E. T. (2011). Nociceptive behavior and physiology of molluscs: Animal welfare implications. *ILAR Journal, 52*(2), 185–195.

Crump, A., Browning, H., Schnell, A., Burn, C., & Birch, J. (2022a). Animal sentience research: Synthesis and proposals. *Animal Sentience, 7*(32), 31.

Crump, A., Browning, H., Schnell, A., Burn, C., & Birch, J. (2022b). Sentience in decapod crustaceans: A general framework and review of the evidence. *Animal Sentience, 7*(32), 1.

Damasio, A. R. (2000). *The feeling of what happens: Body and emotion in the making of consciousness.* Random House.

Damasio, A. (2010). *Self comes to mind: Constructing the conscious brain.* Vintage.

Damasio, A., & Carvalho, G. B. (2013). The nature of feelings: Evolutionary and neurobiological origins. *Nature Reviews Neuroscience, 14,* 143–152.

Dardis A. (2008). *Mental causation: The mind–body problem.* Columbia University Press.

Davies, P. (2006). Preface. In P. Clayton & P. Davies (Eds.), *The re-emergence of emergence,* Ed. P. Clayton & P, Davies, (pp. ix–xix). Oxford: Oxford University Press.

Deacon, T. W. (2011). *Incomplete nature: How mind emerged from matter.* WW Norton & Company.

Dekkers, M. P., Salfelder, F., Sanders, T., Umuerri, O., Cohen, N., & Jansen, G. (2021). Plasticity in gustatory and nociceptive neurons controls decision making in *C. elegans* salt navigation. *bioRxiv.*

Dennett, D. C. (1988). Quining qualia. In A. J. Marcel & E. Bisiach (Eds.), *Consciousness in contemporary science.* Clarendon Press.

Denton, D. (2006). *The primordial emotions: The dawning of consciousness*. Oxford University Press.

Dexter, J. P., Prabakaran, S., & Gunawardena, J. (2019). A complex hierarchy of avoidance behaviors in a single-cell eukaryote. *Current Biology, 29*(24), 4323–4329.

Doerig, A., Schurger, A., & Herzog, M. H. (2021). Hard criteria for empirical theories of consciousness. *Cognitive Neuroscience, 12*(2), 41–62.

D'Silva, J., & Turner, J. (Eds.). (2006). *Animals, ethics and trade: The challenge of animal sentience*. Routledge.

Dussutour, A. (2021). Learning in single-celled organisms. *Biochemical and Biophysical Research Communications, 564*, 92–102.

Edelman, G. M. (1992). *Bright air, brilliant fire: On the matter of the mind*. Basic Books.

Edelman, D. B., & Seth, A. K. (2009). Animal consciousness: A synthetic approach. *Trends in Neurosciences, 32*(9), 476–484.

Egbert, Matthew D., Barandiaran, Xabier E., & Di Paolo, Ezequiel A. (2010). A minimal model of metabolism-based chemotaxis. *PLOS Computational Biology, 6*(12), e1001004.

Eichler, K., Li, F., Litwin-Kumar, A., Park, Y., Andrade, I., Schneider-Mizell, C. M., & Cardona, A. (2017). The complete connectome of a learning and memory centre in an insect brain. *Nature, 548*(7666), 175–182.

Ellia, F., & Chis-Ciure, R. (2022). Consciousness and complexity: Neurobiological naturalism and integrated information theory. *Consciousness and Cognition, 100*, 103281.

Ellia, F., Hendren, J., Grasso. M., Kozma, C., Mindt, G., Lang, J. P., Haun A. M., Albantakis, L., Boly, M., & Tononi, G. (2021). Consciousness and the fallacy of misplaced objectivity. *Neuroscience of Consciousness, 7*(2), 1–12.

Elwood, R. W. (2019). Discrimination between nociceptive reflexes and more complex responses consistent with pain in crustaceans. *Philosophical Transactions of the Royal Society B, 374*(1785), 20190368.

Elwood, R. W. (2022). Hermit crabs, shells, and sentience. *Animal Cognition, 25*(5), 1241–1257.

Emmeche, C., Køppe, S., & Stjernfelt, F. (2000). Levels, emergence, and three versions of downward causation. In P. B. Andersen, C. Emmeche, N. O. Finnemann & P. V. Christiansen (Eds.), *Downward causation: Minds, bodies and matter* (pp. 13–34). Aarhus University Press.

Eronen, M. I. (2021). Levels, nests and branches: Compositional organization and downward causation in biology. In D. S. Brooks, J. DiFrisco, & W.C Wimsatt (Eds.), *Levels of organization in the biological sciences* (pp. 77–87). MIT Press.

Eschbach, C., Fushiki, A., Winding, M., Afonso, B., Andrade, I. V., Cocanougher, B. T., Eichler, K., Gepner, R., Si, G., Valdes-Aleman, J., Fetter, R. D., Gershow, M., Jefferis, G. S., Samuel, A. D., Truman, J. W., Cardona, A., & Zlatic, M. (2021). Circuits for integrating learned and innate valences in the insect brain. *eLife, 10*, e62567.

Evans, S. D., Hughes, I. V., Gehling, J. G., & Droser, M. L. (2020). Discovery of the oldest bilaterian from the Ediacaran of South Australia. *Proceedings of the National Academy of Sciences, 117*(14), 7845–7850.

Fallon, F. (2016). Integrated information theory of consciousness. *Internet encyclopedia of philosophy.* from https://iep.utm.edu/int-info/

Fang-Yen, C., Alkema, M. J., & Samuel, A. D. (2015). Illuminating neural circuits and behaviour in Caenorhabditis elegans with optogenetics. *Philosophical Transactions of the Royal Society B: Biological Sciences, 370*(1677), 20140212.

Fanjul, E., Grela, M. A., Canepuccia, A., & Iribarne, O. (2008). The Southwest Atlantic intertidal burrowing crab *Neohelice granulata* modifies nutrient loads of phreatic waters entering coastal area. *Estuarine, Coastal and Shelf Science, 79*(2), 300–306.

Feigl, H. (1958). *The "mental" and the "physical."* University of Minnesota Press.

Feinberg, T. E. (2001). Why the mind is not a radically emergent feature of the brain. *Journal of Consciousness Studies, 8*, 123–145.

Feinberg, T. E. (2012). Neuroontology, Neurobiological naturalism, and consciousness: A challenge to scientific reduction and a solution. *Physics of Life Reviews, 9*(1), 13–34.

Feinberg, T. E., & Mallatt, J. (2013). The evolutionary and genetic origins of consciousness in the Cambrian period over 500 million years ago. *Frontiers in Psychology, 4*, 667.

Feinberg, T. E., & Mallatt, J. (2016a). *The ancient origins of consciousness: How the brain created experience.* MIT Press.

Feinberg, T. E., & Mallatt, J. (2016b). The nature of primary consciousness: A new synthesis. *Consciousness and Cognition, 43*, 113–127.

Feinberg, T. E., & Mallatt, J. (2018a). *Consciousness demystified.* MIT Press.

Feinberg, T. E., & Mallatt, J. (2018b). Unlocking the "mystery" of consciousness. *Scientific American: Observations.* https://blogs.scientificamerican.com/observations /unlocking-the-mystery-of-consciousness

Feinberg, T. E., & Mallatt, J. (2019). Subjectivity "demystified": Neurobiology, evolution, and the explanatory gap. *Frontiers in Psychology, 10*, 1686.

Feinberg, T. E., & Mallatt, J. (2020). Phenomenal consciousness and emergence: Eliminating the explanatory gap. *Frontiers in Psychology, 11*, 1041.

Feinstein, J. S. (2013). Lesion studies of human emotion and feeling. *Current Opinion in Neurobiology, 23*(3), 304–309.

Finn, J. K., Tregenza, T., & Norman, M. D. (2009). Defensive tool use in a coconut-carrying octopus. *Current Biology, 19*(23), R1069–R1070.

Fischer, D. B., & Truog, R. D. (2015). What is a reflex?: A guide for understanding disorders of consciousness. *Neurology, 85*(6), 543–548.

Flanagan, O. (1992). *Consciousness reconsidered.* MIT Press.

Frazer, J. (2021). Can a cell make decisions? *Scientific American.* https://www.scientificamerican.com/article/can-a-cell-make-decisions

Gans, C., & Northcutt, R. G. (1983). Neural crest and the origin of vertebrates: A new head. *Science, 220*(4594), 268–273.

Ghosh, D. D., Nitabach, M. N., Zhang, Y., & Harris, G. (2017). Multisensory integration in *C. elegans. Current Opinion in Neurobiology, 43*, 110–118.

Gibb, S., Hendry, R. F., & Lancaster, T. (Eds.). (2019). *The Routledge handbook of emergence.* Routledge.

Gibbons, M., Crump, A., Barrett, M., Sarlak, S., Birch, J., & Chittka, L. (2022a). Can insects feel pain? A review of the neural and behavioural evidence. *Advances in Insect Physiology, 63*, 155–229.

Gibbons, M., Sarlak, S., & Chittka, L. (2022b). Descending control of nociception in insects? *Proceedings of the Royal Society B, 289*(1978), 20220599.

Gibbons, M., Versace, E., Crump, A., Baran, B., & Chittka, L. (2022c). Motivational trade-offs and modulation of nociception in bumblebees. *Proceedings of the National Academy of Sciences, 119*, 31, 1–3.

Gillett, C. (2002a). Strong emergence as a defense of non-reductive physicalism. *Principia: An International Journal of Epistemology, 6*(1), 89–120.

Gillett, C. (2002b). The varieties of emergence: Their purposes, obligations and importance. *Grazer Philosophische Studien, 65*(1), 95–121.

Gillette, R., & Brown, J.W. (2015). The sea slug, *Pleurobranchaea californica*: A signpost species in the evolution of complex nervous systems and behavior. *Integrative and Comparative Biology, 55*(6), 1–12.

Ginsburg, S., & Jablonka, E. (2019). *The evolution of the sensitive soul: Learning and the origins of consciousness.* MIT Press.

Giurfa, M. (2013). Cognition with few neurons: Higher-order learning in insects. *Trends in Neurosciences, 36*(5), 285–294.

Glauser, D. A., Chen, W. C., Agin, R., et al. (2011). Heat avoidance is regulated by transient receptor potential (TRP) channels and a neuropeptide signaling pathway in Caenorhabditis elegans. *Genetics, 188*(1), 91–103.

Glover, J. C., & Fritzsch, B. (2009). Brains of primitive chordates. *Encyclopedia of Neurosciences*, 439–448.

Godfrey-Smith, P. (2016). *Other minds: The octopus, the sea, and the deep origins of consciousness*. Farrar, Straus and Giroux.

Godfrey-Smith, P. (2019a). Evolving across the explanatory gap. *Philosophy, Theory, and Practice in Biology, 11*(1), 1–24.

Godfrey-Smith, P. (2019b). Octopus experience. *Animal Sentience, 4*(26), 18.

Goff, P., Seager, W., & Allen-Hermanson, S. (2021). Panpsychism. *The Stanford encyclopedia of philosophy* (Winter 2021 Edition, Edward N. Zalta, ed.). https://plato.stanford.edu/archives/win2021/entries/panpsychism/

Goldstein, J. A. (2018). Emergence and radical novelty: From theory to methods. In E. Mitleton-Kelly, A. Paraskevas, & C. Day (Eds.), *Handbook of research methods in complexity science* (pp. 507–524). Edward Elgar Publishing.

Goodson, J. L., & Kingsbury, M. A. (2013). What's in a name? Considerations of homologies and nomenclature for vertebrate social behavior networks. *Hormones and Behavior, 64*(1), 103–112.

Grasso, F. W. (2014). The octopus with two brains: How are distributed and central representations integrated in the octopus central nervous system? In A. Darmaillacq, L. Dickel, & J. Mather (Eds.), *Cephalopod cognition* (pp. 94–122). Cambridge University Press.

Graziadei, P. (1964). Electron microscopy of some primary receptors in the sucker of Octopus vulgaris. *Zeitschrift für Zellforschung und mikroskopische Anatomie, 64*(4), 510–522.

Grillner, S., & El Manira, A. (2020). Current principles of motor control, with special reference to vertebrate locomotion. *Physiological Reviews, 100*(1), 271–320.

Gunawardena, J. (2022). Learning outside the brain: Integrating cognitive science and systems biology. *Proceedings of the IEEE, 10*(5), 590–612

Gutnick, T., Byrne, R. A., Hochner, B., & Kuba, M. (2011). *Octopus vulgaris* uses visual information to determine the location of its arm. *Current Biology, 21*(6), 460–462.

Haeckel, E. (1892). Our monism: The principles of a consistent, unitary world-view. *The Monist, 2*(4), 481–486.

Hall, B. K. (2008). *The neural crest and neural crest cells in vertebrate development and evolution* (Vol. 11). Springer Science & Business Media.

Hallman, C., & French, K. (2015). Eukaryotes: A new timetable of evolution. *Max-Planck-Gessellschaft*. https://www.mpg.de/9256248/eukaryotes-evolution

Hanlon, R. T., & Messenger, J. B. (2018). *Cephalopod behaviour*. Cambridge University Press.

Hanlon, R. T., Conroy, L. A., & Forsythe, J. W. (2008). Mimicry and foraging behaviour of two tropical sand-flat octopus species off North Sulawesi, Indonesia. *Biological Journal of the Linnean Society, 93*(1), 23–38.

Hanlon, R. T., Forsythe, J. W., & Joneschild, D. (1999). Crypsis, conspicuousness, mimicry and polyphenism as antipredator defences of foraging octopuses on Indo-Pacific coral reefs, with a method of quantifying crypsis from video tapes. *Biological Journal of the Linnean Society, 66*(1), 1–22.

Hanlon, R. T., Watson, A. C., & Barbosa, A. (2010). A "mimic octopus" in the Atlantic: Flatfish mimicry and camouflage by *Macrotritopus defilippi*. *Biological Bulletin, 218*(1), 15–24.

Hansen, T. F. (2015). *Evolutionary constraints*. Oxford University Press.

Harnad, S. (2016). Animal sentience: The other-minds problem. *Animal Sentience, 1*(1), 1–10.

Harnad, S. (2021). On the (too) many faces of consciousness. *Journal of Consciousness Studies, 28*(7–8), 61–66.

Harzsch, S., & Krieger, J. (2018). Crustacean olfactory systems: A comparative review and a crustacean perspective on olfaction in insects. *Progress in Neurobiology, 161*, 23–60.

Hasan, A., & Fumerton, R. (2019). Knowledge by acquaintance vs. description. *The Stanford encyclopedia of philosophy*. https://plato.stanford.edu/archives/sum2019/entries/knowledge-acquaindescrip/

Haswell, E. S., Phillips, R., & Rees, D. C. (2011). Mechanosensitive channels: What can they do and how do they do it? *Structure, 19*(10), 1356–1369.

Hattori, D., Aso, Y., Swartz, K. J., Rubin, G. M., Abbott, L. F., & Axel, R. (2017). Representations of novelty and familiarity in a mushroom body compartment. *Cell, 169*, 956–969.

Heil, J. (2019). Emergence and Panpsychism. In S. Gibb, R. F. Hendry, & T. Lancaster (Eds.), *The Routledge handbook of emergence* (pp. 225–234). Routledge.

Heil, J., & Mele, A. (1993). *Mental causation*. Clarendon Press.

Heisenberg, M., Wolf, R., & Brembs, B. (2001). Flexibility in a single behavioral variable of Drosophila. *Learning and Memory, 8*(1), 1–10.

Hilgetag, C. C., & Goulas, A. (2020). 'Hierarchy' in the organization of brain networks. *Philosophical Transactions of the Royal Society B, 375*(1796), 20190319.

Hilgetag, C. C., & Hütt, M. T. (2014). Hierarchical modular brain connectivity is a stretch for criticality. *Trends in Cognitive Sciences, 18*(3), 114–115.

Hemmi, J. M. (2005). Predator avoidance in fiddler crabs: 2. The visual cues. *Animal Behaviour, 69*(3), 615–625.

Hemmi, J. M., & Tomsic, D. (2012). The neuroethology of escape in crabs: From sensory ecology to neurons and back. *Current Opinion in Neurobiology, 22*(2), 194–200.

Herberholz, J., & Marquart, G. D. (2012). Decision making and behavioral choice during predator avoidance. *Frontiers in Neuroscience, 6,* 125.

Hirayama, K., & Gillette, R. (2012). A neuronal network switch for approach/avoidance toggled by appetitive state. *Current Biology, 22*(2), 118–123.

Hirayama, K., Catanho, M., Brown, J. W., & Gillette, R. (2012). A core circuit module for cost/benefit decision. *Frontiers in Neuroscience, 6,* 123.

Hirayama, K., Moroz, L. L., Hatcher, N. G., & Gillette, R. (2014). Neuromodulatory control of a goal-directed decision. *PLoS One, 9*(7), e102240.

Hochner, B. (2008). Octopuses. *Current Biology, 18*(19), R897–R898.

Hochner, B. (2012). An embodied view of octopus neurobiology. *Current Biology, 22*(20), R887–R892.

Hochner, B., Shomrat, T., & Fiorito, G. (2006). The octopus: A model for a comparative analysis of the evolution of learning and memory mechanisms. *Biological Bulletin, 210*(3), 308–317.

Holland, L. Z. (2015). Genomics, evolution and development of amphioxus and tunicates: The Goldilocks principle. *Journal of Experimental Zoology Part B: Molecular and Developmental Evolution, 324*(4), 342–352.

Holland, N. D., & Chen, J. (2001). Origin and early evolution of the vertebrates: New insights from advances in molecular biology, anatomy, and palaeontology. *Bioessays, 23*(2), 142–151.

Holland, N. D., & Yu, J. K. (2002). Epidermal receptor development and sensory pathways in vitally stained amphioxus (*Branchiostoma floridae*). *Acta Zoologica, 83*(4), 309–319.

Horgan, J. (2015). Can integrated information theory explain consciousness? *Scientific American.* https://blogs.scientificamerican.com/cross-check/can-integrated -information-theory-explain-consciousness/

Hu, H. (2016). Reward and aversion. *Annual Review of Neuroscience, 39,* 297–324.

Hu, W. T., Peng, Y. Q., Sun, J. M., Zhang, F., Zhang, X. C., Wang, L. Z., et al. (2018). Fan-shaped body neurons in the Drosophila brain regulate both innate and conditioned nociceptive avoidance. *Cell Reports, 24,* 1573–1584.

Huffard, C. L. (2006). Locomotion by *Abdopus aculeatus* (Cephalopoda: Octopodidae): Walking the line between primary and secondary defenses. *Journal of Experimental Biology, 209*(19), 3697–3707.

Hultborn, H. (2006). Spinal reflexes, mechanisms and concepts: From Eccles to Lundberg and beyond. *Progress in Neurobiology, 78*(3–5), 215–232.

Ikemoto, S. (2010). Brain reward circuitry beyond the mesolimbic dopamine system: A neurobiological theory. *Neuroscience & Biobehavioral Reviews, 35*(2), 129–150.

Illich, P. A., & Walters, E. T. (1997). Mechanosensory neurons innervating Aplysia siphon encode noxious stimuli and display nociceptive sensitization. *Journal of Neuroscience, 17*(1), 459–469.

Iliff, A. J., & Xu, X. S. (2020). C. elegans: A sensible model for sensory biology. *Journal of Neurogenetics, 34*(3–4), 347–350.

Irwin, L. N. (2020). Renewed perspectives on the deep roots and broad distribution of animal consciousness. *Frontiers in Systems Neuroscience, 14*, 57.

Iwano, M., Hill, E. S., Mori, A., Mishima, T., Mishima, T., Ito, K., & Kanzaki, R. (2010). Neurons associated with the flip-flop activity in the lateral accessory lobe and ventral protocerebrum of the silkworm moth brain. *Journal of Comparative Neurology, 518*(3), 366–388.

Jablonka, E., & Ginsburg, S. (2022). Pain sentience criteria and their grading. *Animal Sentience, 32*(4), 1–4.

Jackson, F. (1982). Epiphenomenal qualia. *Philosophical Quarterly, 32*, 127–136.

Jackson, F. (1986). What Mary didn't know. *The Journal of Philosophy, 83*, 291–295.

Jackson, F. (2004). Forward. Looking back on the knowledge argument. In P. Ludlow, Y. Nagasawa, & D. Stoljar (Eds.), *There's something about Mary: Essays on phenomenal consciousness and Frank Jackson's knowledge argument* (pp. xv–xix). MIT Press.

Jackson, R. R., & Wilcox, R. S. (1993). Observations in nature of detouring behaviour by *Portia fimbriata*, a web-invading aggressive mimic jumping spider from Queensland. *Journal of Zoology, 230*(1), 135–139.

Jackson, R. R., & Wilcox, R. S. (1998). Spider-eating spiders: Despite the small size of their brain, jumping spiders in the genus *Portia* outwit other spiders with hunting techniques that include trial and error. *American Scientist, 86*, 350–357.

Jacobson, G. A., & Friedrich, R. W. (2013). Neural circuits: Random design of a higher-order olfactory projection. *Current Biology, 23*(10), R448–R451.

James, W. (1904). Does consciousness exist? *The Journal of Philosophy, Psychology and Scientific Methods, 1*(18), 477–491.

Jenkins, K. M., Briggs, D. E., & Luque, J. (2022). The remarkable visual system of a Cretaceous crab. *Iscience, 25*(1), 103579.

Jennings, H. S. (1906). *Behavior of the lower organisms*. Columbia University Press.

Jing, J., & Gillette, R. (2000). Escape swim network interneurons have diverse roles in behavioral switching and putative arousal in Pleurobranchaea. *Journal of Neurophysiology, 83*(3), 1346–1355.

Kaas, J. H. (1997). Topographic maps are fundamental to sensory processing. *Brain Research Bulletin, 44*(2), 107–112.

Kaifu, K., Segawa, S., & Tsuchiya, K. (2007). Behavioral responses to underwater sound in the small benthic octopus *Octopus ocellatus. The Journal of the Marine Acoustics Society of Japan, 34*(4), 266–273.

Kaiser, M., Goerner, M., & Hilgetag, C. C. (2007). Criticality of spreading dynamics in hierarchical cluster networks without inhibition. *New Journal of Physics, 9*(5), 110.

Kaiser, M., & Hilgetag, C. C. (2010). Optimal hierarchical modular topologies for producing limited sustained activation of neural networks. *Frontiers in Neuroinformatics, 4*, 8.

Kaiser, M., Hilgetag, C. C., & Kötter, R. (2010). Hierarchy and dynamics of neural networks. *Frontiers in Neuroinformatics, 4*, 112.

Kandel E. R., Koester J. D., Mack S. H., & Siegelbaum S. A. (2021). *Principles of neural science*. 6th ed. McGraw-Hill Education.

Katsuki, T., & Greenspan, R. J. (2013). Jellyfish nervous systems. *Current Biology, 23*(14), R592–R594.

Kearns, D. B. (2010). A field guide to bacterial swarming motility. *Nature Reviews Microbiology, 8*(9), 634–644.

Kender, R. G., Harte, S. E., Munn, E. M., & Borszcz, G. S. (2008). Affective analgesia following muscarinic activation of the ventral tegmental area in rats. *The Journal of Pain, 9*(7), 597–605.

Key, B., Zalucki, O., & Brown, D. J. (2021). Neural design principles for subjective experience: Implications for insects. *Frontiers in Behavioral Neuroscience, 15*, 658037.

Keymer, J. E., Endres, R. G., Skoge, M., Meir, Y., & Wingreen, N. S. (2006). Chemosensing in Escherichia coli: Two regimes of two-state receptors. *Proceedings of the National Academy of Sciences, 103*(6), 1786–1791.

Kim, J. (1992). "Downward causation" in emergentism and nonreductive physicalism. In A. Beckermann, H. Flohr, & J. Kim (Eds.), *Emergence or reduction?: Essays on the prospects of nonreductive physicalism* (pp. 119–138). De Gruyter.

Kim, J. (1998). *Mind in a physical world. An essay on the mind–body problem and mental causation.* MIT Press.

Kim, J. (2000). Making sense of downward causation. In P. B. Andersen, C. Emmeche, N. O. Finnemann, & P. V. Christiansen (Eds.), *Downward causation* (pp. 305–321). Aarhus University Press.

Kim J. (2006). Being realistic about emergence. In P. Clayton & P. Davies (Eds.), *The re-emergence of emergence* (pp. 190–202). Oxford University Press.

Kirwan, J. D., Graf, J., Smolka, J., Mayer, G., Henze, M. J., & Nilsson, D. E. (2018). Low-resolution vision in a velvet worm (Onychophora). *Journal of Experimental Biology, 221*(11), jeb175802.

Klein, C., & Barron, A. B. (2016). Insects have the capacity for subjective experience. *Animal Sentience, 1*(9), 1.

Kobayashi, S., Kojima, S., Yamanaka, M., Sadamoto, H., Nakamura, H., Fujito, Y., & Ito, E. (1998). Operant conditioning of escape behavior in the pond snail, *Lymnaea stagnalis. Zoological Science, 15*(5), 683–690.

Koch, C. (2012). *Consciousness: Confessions of a romantic reductionist.* MIT Press.

Koch, C. (2019). *The feeling of life itself: Why consciousness is widespread but can't be computed.* MIT Press.

Koch, C., & Laurent, G. (1999). Complexity and the nervous system. *Science, 284*(5411), 96–98.

Koch, C., Massimini, M., Boly, M., & Tononi, G. (2016). Neural correlates of consciousness: Progress and problems. *Nature Reviews Neuroscience, 17*(5), 307–321.

Koester, J., Siegelbaum, S. A., Kandel, E. R., Schwartz, J. H., & Jessell, T. M. (2021). *Principles of neural science.* 6th ed. McGraw Hill.

Krieger, J., Braun, P., Rivera, N. T., Schubart, C. D., Müller, C. H., & Harzsch, S. (2015). Comparative analyses of olfactory systems in terrestrial crabs (Brachyura): Evidence for aerial olfaction? *PeerJ, 3*, e1433.

Krieger, J., Sombke, A., Seefluth, F., Kenning, M., Hansson, B. S., & Harzsch, S. (2012). Comparative brain architecture of the European shore crab *Carcinus maenas* (Brachyura) and the common hermit crab *Pagurus bernhardus* (Anomura) with notes on other marine hermit crabs. *Cell and Tissue Research, 348*(1), 47–69.

Kringelbach, M. L., & Berridge, K. C. (2017). The affective core of emotion: Linking pleasure, subjective well-being, and optimal metastability in the brain. *Emotion Review, 9*(3), 191–199.

Kumazawa, H. (2002). Notes on the taxonomy of Stentor Oken (Protozoa, Ciliophora) and a description of a new species. *Journal of Plankton Research, 24*(1), 69–75.

Lacalli, T. C. (1996). Frontal eye circuitry, rostral sensory pathways and brain organization in amphioxus larvae: Evidence from 3D reconstructions. *Philosophical Transactions of the Royal Society of London. Series B: Biological Sciences, 351*(1337), 243–263.

Lacalli, T. C. (2004). Sensory systems in amphioxus: A window on the ancestral chordate condition. *Brain, Behavior and Evolution, 64*(3), 148–162.

Lacalli, T. C. (2008). Basic features of the ancestral chordate brain: A protochordate perspective. *Brain Research Bulletin, 75*(2–4), 319–323.

Lacalli, T. C. (2010). The emergence of the chordate body plan: Some puzzles and problems. *Acta Zoologica, 91*(1), 4–10.

Lacalli, T. (2018a). Amphioxus, motion detection, and the evolutionary origin of the vertebrate retinotectal map. *EvoDevo, 9*(1), 1–5.

Lacalli, T. (2018b). Amphioxus neurocircuits, enhanced arousal, and the origin of vertebrate consciousness. *Consciousness and Cognition, 62*, 127–134.

Lacalli, T. (2021). Innovation through heterochrony: An amphioxus perspective on telencephalon origin and function. *Frontiers in Ecology and Evolution, 9*, 318.

Lacalli, T. (2022). An evolutionary perspective on chordate brain organization and function: Insights from amphioxus, and the problem of sentience. *Philosophical Transactions of the Royal Society B, 377*(1844), 20200520.

Lacalli, T., & Candiani, S. (2017). Locomotory control in amphioxus larvae: New insights from neurotransmitter data. *EvoDevo, 8*(1), 1–8.

Ladyman, J., & Wiesner, K. (2020). *What is a complex system?* Yale University Press.

Laidre, M. E. (2014). The social lives of hermits. *Natural History, 122*, 24–29.

Laidre, M. E. (2021). The architecture of cooperation among non-kin: Coalitions to move up in nature's housing market. *Frontiers in Ecology and Evolution, 9*, 766342.

Lamme, V. A. (2006). Towards a true neural stance on consciousness. *Trends in Cognitive Sciences, 10*(11), 494–501..

Levandowsky, M., Cheng, T., Kehr, A., Kim, J., Gardner, L., Silvern, L., & Prakash, E. (1984). Chemosensory responses to amino acids and certain amines by the ciliate Tetrahymena: A flat capillary assay. *Biological Bulletin, 167*(2), 322–330.

Levine, J. (1983). Materialism and phenomenal properties: The explanatory gap. *Pacific Philosophical Quarterly, 64*(4), 354–361.

Lewes, G. H. (1877). *Problems of life and mind.* Trübner & Company.

Lewin, R. (1992). *Complexity: Life at the edge of chaos.* University of Chicago Press.

Lewis, R. (2020). The evolutionary origins of consciousness. Part 2: The ancient beginnings of subjective experience. *Psychology Today.* https://www.psychologytoday.com/us/blog/finding-purpose/202010/the-evolutionary-origins-consciousness

Loukola, O. J., Solvi, C., Coscos, L., & Chittka, L. (2017). Bumblebees show cognitive flexibility by improving on an observed complex behavior. *Science, 355*(6327), 833–836.

Ludlow, P., Nagasawa, Y., & Stoljar, D. (Eds.). (2004). *There's something about Mary: Essays on phenomenal consciousness and Frank Jackson's knowledge argument.* MIT Press.

Lyon, P. (2015). The cognitive cell: Bacterial behavior reconsidered. *Frontiers in Microbiology, 6,* 264.

Macat Team. (2017). *An analysis of William James's the principles of psychology.* CRC Press.

Macdonald, C., & Macdonald, G. (Eds.). (2010). *Emergence in mind.* Oxford University Press.

Macnab, R. M. (1999). The bacterial flagellum: Reversible rotary propellor and type III export apparatus. *Journal of Bacteriology, 181*(23), 7149–7153.

Maldonado, H. (1965). The positive and negative learning process in *Octopus vulgaris* Lamarck. Influence of the vertical and median superior frontal lobes. *Zeitschrift für vergleichende Physiologie, 51,* 185–203.

Mallatt, J. (2021). A traditional scientific perspective on the integrated information theory of consciousness. *Entropy, 23*(6), 650.

Mallatt, J., & Feinberg, T. E. (2017). Consciousness is not inherent in but emergent from life. *Animal Sentience, 11*(15), 1–7.

Mallatt, J., & Feinberg, T. E. (2020). Sentience in evolutionary context. *Animal Sentience, 5*(29), 14.

Mallatt, J., & Feinberg, T. E. (2021). Multiple routes to animal consciousness: Constrained multiple realizability rather than modest identity theory. *Frontiers in Psychology, 12.* https://doi.org/10.3389/fpsyg.2021.732336

Mamlouk, A. M., & Schmuker, M. (2011). Self-organization of virtual odorant receptors inspired by insect olfaction. *BMC Neuroscience, 12*(1), 1–2.

Mancini, F., Haggard, P., Iannetti, G. D., Longo, M. R., & Sereno, M. I. (2012). Fine-grained nociceptive maps in primary somatosensory cortex. *Journal of Neuroscience, 32*(48), 17155–17162.

Marini, G., De Sio, F., Ponte, G., & Fiorito, G. (2017). Behavioral analysis of learning and memory in cephalopods. In J. H. Byrne (Ed.), *Learning and memory: A comprehensive reference* (2nd ed.) (pp. 441–462). Academic Press.

Martin, C., Jahn, H., Klein, M., Hammel, J. U., Stevenson, P. A., Homberg, U., & Mayer, G. (2022). The velvet worm brain unveils homologies and evolutionary novelties across panarthropods. *BMC Biology, 20*(1), 1–51.

Mather, J. (2012). Cephalopod intelligence. In J. Vonk & T. K. Shackelford (Eds.), *The Oxford handbook of comparative evolutionary psychology* (pp. 118–128). Oxford University Press.

Mather, J. (2019). What is in an octopus's mind? *Animal Sentience, 26(1),* 1–29.

Mather, J. (2021a). The case for octopus consciousness: Unity. *NeuroSci, 2*(4), 405–415.

Mather, J. (2021b). Octopus consciousness: The role of perceptual richness. *NeuroSci, 2*(3), 276–290.

Mather, J. A. (1995). Cognition in cephalopods. *Advances in the Study of Behaviour, 24,* 317.

Mather, J. A. (2008). Cephalopod consciousness: Behavioural evidence. *Consciousness and Cognition, 17*(1), 37–48.

Mayer, G., Whitington, P. M., Sunnucks, P., & Pflüger, H. J. (2010). A revision of brain composition in Onychophorans (velvet worms) suggests that the tritocerebrum evolved in arthropods. *BMC Evolutionary Biology, 10*(1), 1–9.

Mayr, E. (1982). *The growth of biological thought: Diversity, evolution, and inheritance.* Harvard University Press.

Mayr, E. (2004). *What makes biology unique? Considerations on the autonomy of a scientific discipline.* Cambridge University Press.

Medan, V., Oliva, D., & Tomsic, D. (2007). Characterization of lobula giant neurons responsive to visual stimuli that elicit escape behaviors in the crab Chasmagnathus. *Journal of Neurophysiology, 98*(4), 2414–2428.

Meehl, P. (1966). The compleat autocerebroscopist: A thought experiment on Professor Feigl's mind/body identify thesis. In P. K. Feyerabend & G. Maxwell (Eds.), *Mind, matter, and method* (pp. 103–180). University of Minnesota Press.

Merker, B., Williford, K., & Rudrauf, D. (2022). The integrated information theory of consciousness: A case of mistaken identity. *Behavioral and Brain Sciences, 45,* e41, 1–63.

Messenger, J. B. (2001). Cephalopod chromatophores: Neurobiology and natural history. *Biological Reviews, 76*(4), 473–528.

Metaxakis, A., Petratou, D., & Tavernarakis, N. (2018). Multimodal sensory processing in Caenorhabditis elegans. *Open Biology, 8*(6), 180049.

Metzinger, T. (2003). *Being no one: The self-model theory of subjectivity.* MIT Press.

Meunier, D., Lambiotte, R., Fornito, A., Ersche, K., & Bullmore, E. T. (2009). Hierarchical modularity in human brain functional networks. *Frontiers in Neuroinformatics, 3*, 37.

Micali, G., & Endres, R. G. (2016). Bacterial chemotaxis: Information processing, thermodynamics, and behavior. *Current Opinion in Microbiology, 30*, 8–15.

Michel, M., Beck, D., Block, N., Blumenfeld, H., Brown, R., Carmel, D., Carrasco, M., Chirimuuta, M., Chun, M., & Cleeremans, (2019). A. Opportunities and challenges for a maturing science of consciousness. *Nature Human Behavior, 3*(2), 104–107.

Mikhalevich, I., & Powell, R. (2020). Minds without spines: Evolutionarily inclusive animal ethics. *Animal Sentience, 29*(1), 1–25.

Mills, H., Ortega, A., Law, W., Hapiak, V., Summers, P., Clark, T., & Komuniecki, R. (2016). Opiates modulate noxious chemical nociception through a complex monoaminergic/peptidergic cascade. *Journal of Neuroscience, 36*(20), 5498–5508.

Min, B. K. (2010). A thalamic reticular networking model of consciousness. *Theoretical Biology and Medical Modelling, 7*(1), 1–18.

Mitchell, M. (2009). *Complexity: A guided tour.* Oxford University Press.

Montgomery, S. (2015). *The soul of an octopus: A surprising exploration into the wonder of consciousness.* Simon and Schuster.

Morowitz, H. J. (2002). *The emergence of everything: how the world became complex.* Oxford University Press.

Moroz, L. L. (2011). Aplysia. *Current Biology: CB, 21*(2), R60.

Moroz, L. L., Edwards, J. R., Puthanveettil, S. V., Kohn, A. B., Ha, T., Heyland, A., & Kandel, E. R. (2006). Neuronal transcriptome of Aplysia: Neuronal compartments and circuitry. *Cell, 127*(7), 1453–1467.

Moroz, L. L., Hatcher, N. G., & Gillette, R. (2014). Neuromodulatory control of a goal-directed decision. *PLoS ONE, 9*(7), e102240.

Moynihan, M., & Rodaniche, A. F. (1982). The behavior and natural history of the Caribbean reef squid *Sepioteuthis sepioidea* with a consideration of social, signal, and defensive patterns for difficult and dangerous environments. *Advanced Ethology, 25*, 9–150.

Müller, W. E. G., Li, J., Schröder, H. C., Qiao, L., & Wang, X. (2007). The unique skeleton of siliceous sponges (Porifera; Hexactinellida and Demospongiae) that evolved first from the Urmetazoa during the Proterozoic: A review. *Biogeosciences, 4*(2), 219–232.

Nagel, T. (1974). What is it like to be a bat? *The Philosophical Review, 83*(4), 435–450.

Nagel, T. (1986). *The view from nowhere.* Oxford University Press.

Nagel, T. (1995). *Other minds: Critical essays, 1969–1994*. Oxford University Press on Demand.

Newland, P. L., Rogers, S. M., Gaaboub, I., & Matheson, T. (2000). Parallel somatotopic maps of gustatory and mechanosensory neurons in the central nervous system of an insect. *Journal of Comparative Neurology, 425*(1), 82–96.

Nicol, D., & Meinertzhagen, I. A. (1991). Cell counts and maps in the larval central nervous system of the ascidian *Ciona intestinalis* (L.). *Journal of Comparative Neurology, 309*(4), 415–429.

Nida-Rümelin, M., and O Conaill, D. (2019). Qualia: The knowledge argument. In *The Stanford encyclopedia of philosophy*. https://plato.stanford.edu/archives/win2019 /entries/qualia-knowledge/

Nieuwenhuys, R. (1998). Morphogenesis and general structure. In R. Nieuwenhuys, H. J. Ten Donkelaar, & C. Nicholson (Eds.), *The central nervous system of vertebrates* (pp. 159–228). Springer-Verlag.

Nishijima, S., & Maruyama, I. N. (2017). Appetitive olfactory learning and long-term associative memory in *Caenorhabditis elegans*. *Frontiers in Behavioral Neuroscience, 11*, 80.

Nityananda, V., & Chittka, L. (2021). Different effects of reward value and saliency during bumblebee visual search for multiple rewarding targets. *Animal Cognition, 24*(4), 803–814.

Noble, R., Tasaki, K., Noble, P. J., & Noble, D. (2019). Biological relativity requires circular causality but not symmetry of causation: So, where, what and when are the boundaries? *Frontiers in Physiology, 10*, 827.

Noboa, V., & Gillette, R. (2013). Selective prey avoidance learning in the predatory sea-slug *Pleurobranchaea californica*. *Journal of Experimental Biology, 216*, 3231–3236.

Norman, M. D., Finn, J., & Tregenza, T. (1999). Female impersonation as an alternative reproductive strategy in giant cuttlefish. *Proceedings of the Royal Society of London. Series B: Biological Sciences, 266*(1426), 1347–1349.

Norman, M. D., Finn, J., & Tregenza, T. (2001). Dynamic mimicry in an Indo–Malayan octopus. *Proceedings of the Royal Society of London. Series B: Biological Sciences, 268*(1478), 1755–1758.

Northcutt, R. G. (2005). The new head hypothesis revisited. *Journal of Experimental Zoology Part B: Molecular and Developmental Evolution, 304*(4), 274–297.

Nunez, P. L. (2016). *The new science of consciousness*. Prometheus Books.

Oami, K. (1996). Distribution of chemoreceptors to quinine on the cell surface of *Paramecium caudatum*. *Journal of Comparative Physiology A, 179*(3), 345–352.

O'Connell, L. A., & Hofmann, H. A. (2012). Evolution of a vertebrate social decision-making network. *Science, 336*(6085), 1154–1157.

Oizumi, M., Albantakis, L., & Tononi, G. (2014). From the phenomenology to the mechanisms of consciousness: Integrated information theory 3.0. *PLoS Computational Biology, 10*(5), e1003588.

Oliva, D., Medan, V., & Tomsic, D. (2007). Escape behavior and neuronal responses to looming stimuli in the crab *Chasmagnathus granulatus* (Decapoda: Grapsidae). *Journal of Experimental Biology, 210*(5), 865–880.

Overgaard, M. (2021). Insect consciousness. *Frontiers in Behavioral Neuroscience, 15*, 102.

Owald, D., Felsenberg, J., Talbot, C. B., Das, G., Perisse, E., Huetteroth, W., & Waddell, S. (2015). Activity of defined mushroom body output neurons underlies learned olfactory behavior in Drosophila. *Neuron, 86*, 417–427.

Packard, A., & Delafield-Butt, J. T. (2014). Feelings as agents of selection: Putting Charles Darwin back into (extended neo-) Darwinism. *Biological Journal of the Linnean Society, 112*, 332–353.

Panetta, D., Buresch, K., & Hanlon, R. T. (2017). Dynamic masquerade with morphing three-dimensional skin in cuttlefish. *Biology Letters, 13*(3), 20170070.

Panksepp J. (1998). *Affective neuroscience: The foundations of human and animal emotions*. Oxford University Press.

Panksepp J. (2005). Affective consciousness: Core emotional feelings in animals and humans. *Consciousness and Cognition*, 14, 30–80. 10.1016/j.concog.2004.10.004

Papini, M. R., & Bitterman, M. E. (1991). Appetitive conditioning in *Octopus cyanea*. *Journal of Comparative Psychology,105*(2), 107.

Partridge, J. D., Nhu, N. T., Dufour, Y. S., & Harshey, R. M. (2019). *Escherichia coli* remodels the chemotaxis pathway for swarming. *MBio, 10*(2), e00316–e00319.

Pattee, H. H. (1970). The problem of biological hierarchy. In C. H. Waddington (Ed.), *Towards a theoretical biology 3* (pp. 117–136). Aldine.

Patteson, A. E., Gopinath, A., Goulian, M., & Arratia, P. E. (2015). Running and tumbling with E. coli in polymeric solutions. *Scientific Reports, 5*(1), 1–11.

Pergner, J., & Kozmik, Z. (2017). Amphioxus photoreceptors-insights into the evolution of vertebrate opsins, vision and circadian rhythmicity. *International Journal of Developmental Biology, 61*(10–11–12), 665–681.

Perry, C. J., & Chittka, L. (2019). How foresight might support the behavioral flexibility of arthropods. *Current Opinion in Neurobiology, 54*, 171–177.

Peterson, K. J., & Butterfield, N. J. (2005). Origin of the Eumetazoa: Testing ecological predictions of molecular clocks against the Proterozoic fossil record. *Proceedings of the National Academy of Sciences, 102*(27), 9547–9552.

Polese, G., Bertapelle, C., & Di Cosmo, A. (2015). Role of olfaction in *Octopus vulgaris* reproduction. *General and Comparative Endocrinology, 210,* 55–62.

Polese, G., Bertapelle, C., & Di Cosmo, A. (2016). Olfactory organ of *Octopus vulgaris*: Morphology, plasticity, turnover and sensory characterization. *Biology Open, 5*(5), 611–619.

Polger, T. W., & Shapiro, L. A. (2016). *The multiple realization book.* Oxford University Press.

Poncelet, G., & Shimeld, S. M. (2020). The evolutionary origins of the vertebrate olfactory system. *Open Biology, 10*(12), 200330.

Ponder, W. F., & Lindberg, D. R. (2008). *Phylogeny and evolution of the Molluska.* University of California Press.

Ponte, G., Chiandetti, C., Edelman, D. B., Imperadore, P., Pieroni, E. M., & Fiorito, G. (2022). Cephalopod behavior: From neural plasticity to consciousness. *Frontiers in Systems Neuroscience, 15,* 175.

Potochnik, A., & McGill, B. (2012). The limitations of hierarchical organization. *Philosophy of Science, 79*(1), 120–140.

Povich, M., & Craver, C. F. (2017). Mechanistic levels, reduction, and emergence. In S. Glennan & P. Illari (Eds.), *The Routledge handbook of mechanisms and mechanical philosophy* (pp. 185–197). Routledge.

Price, J. L. (2005). Free will versus survival: Brain systems that underlie intrinsic constraints on behavior. *Journal of Comparative Neurology, 493*(1), 132–139.

Pungor, J. R., Allen, V. A., Songco-Casey, J. O., & Niell, C. M. (2023). Functional organization of visual responses in the octopus optic lobe. *Current Biology, 3*(1), 2784–2793.

Putz, G., & Heisenberg, M. (2002). Memories in Drosophila heat-box learning. *Learning and Memory, 9*(5), 349–359.

Reber, A. S. (2019). *The first minds: Caterpillars, karyotes, and consciousness.* Oxford University Press.

Reinhard, J., & Rowell, D. M. (2005). Social behaviour in an Australian velvet worm, *Euperipatoides rowelli* (Onychophora: Peripatopsidae). *Journal of Zoology, 267*(1), 1–7.

Revonsuo, A. (2006). *Inner presence: Consciousness as a biological phenomenon.* MIT Press.

Revonsuo, A. (2010). *Consciousness: The science of subjectivity.* Psychology Press.

Reynierse, J. H., & Walsh, G. L. (1967). Behavior modification in the protozoan Stentor re-examined. *The Psychological Record, 17*(2), 161–165.

Robinson, Howard. Dualism. *The Stanford encyclopedia of philosophy,* Winter 2020 ed. Edward N. Zalta, Ed. https://plato.stanford.edu/archives/fall2021/entries/dualism/.

Rotjan, R. D., Chabot, J. R., & Lewis, S. M. (2010). Social context of shell acquisition in *Coenobita clypeatus* hermit crabs. *Behavioral Ecology, 21*(3), 639–646.

Russell, B. (1910–1911), Knowledge by acquaintance and knowledge by description. *Proceedings of the Aristotelian Society,* 11, 108–128.

Russell, B. (1912). *The problems of philosophy.* Henry Holt and Company.

Russell, B. (1914). On the nature of acquaintance. *Monist,* 24, 161–187.

Sagawa, T., Kikuchi, Y., Inoue, Y., Takahashi, H., Muraoka, T., Kinbara, K., & Fukuoka, H. (2014). Single-cell *E. coli* response to an instantaneously applied chemotactic signal. *Biophysical Journal, 107*(3), 730–739.

Salthe, S. N. (1985). *Evolving hierarchical systems: Their structure and representation.* Columbia University Press.

Scarano, F., & Tomsic, D. (2014). Escape response of the crab Neohelice to computer generated looming and translational visual danger stimuli. *Journal of Physiology-Paris, 108*(2–3), 141–147.

Schafer, W. (2016). Nematode nervous systems. *Current Biology, 26*(20), R955–R959.

Schlosser, G., Patthey, C., & Shimeld, S. M. (2014). The evolutionary history of vertebrate cranial placodes II. Evolution of ectodermal patterning. *Developmental Biology, 389*(1), 98–119.

Schnell, A. K., Amodio, P., Boeckle, M., & Clayton, N. S. (2021). How intelligent is a cephalopod? Lessons from comparative cognition. *Biological Reviews, 96*(1), 162–178.

Schrum, J. P., Zhu T. F., & Szostak, J. W. (2010). The origins of cellular life. *Cold Spring Harbor Perspectives in Biology, 2*(9), a002212.

Searle, J. R. (1992). *The rediscovery of the mind.* MIT Press.

Searle, J. R. (1997). *The mystery of consciousness.* New York Review of Books.

Searle, J. R. (2007). Biological naturalism. In M. Velmans & S. Schneider (Eds.), *The Blackwell companion to consciousness* (pp. 325–334). John Wiley & Sons.

Seelig, J. D., & Jayaraman, V. (2013). Feature detection and orientation tuning in the *Drosophila* central complex. *Nature, 503*, 262–266.

Sellars, W. *Science, perception, and reality.* Routledge and Kegan Paul: 1963.

Shapiro, L. A. (2004). *The mind incarnate*. MIT Press.

Shear, J. (Ed.). (1999). *Explaining consciousness: The hard problem*. MIT Press.

Shepherd, G. M. (2007). Perspectives on olfactory processing, conscious perception, and orbitofrontal cortex. *Annals of the New York Academy of Sciences, 1121*(1), 87–101.

Shigeno, S., & Ragsdale, C. W. (2015). The gyri of the octopus vertical lobe have distinct neurochemical identities. *Journal of Comparative Neurology, 523*(9), 1297–1317.

Shigeno, S., Andrews, P. L., Ponte, G., & Fiorito, G. (2018). Cephalopod brains: An overview of current knowledge to facilitate comparison with vertebrates. *Frontiers in Physiology, 9*, 952.

Shohat-Ophir, G., Kaun, K. R., Azanchi, R., Mohammed, H., & Heberlein, U. (2012). Sexual deprivation increases ethanol intake in *Drosophila*. *Science, 335*(6074), 1351–1355.

Shomrat, T., Turchetti-Maia, A. L., Stern-Mentch, N., Basil, J. A., & Hochner, B. (2015). The vertical lobe of cephalopods: An attractive brain structure for understanding the evolution of advanced learning and memory systems. *Journal of Comparative Physiology A, 201*, 947–956.

Siju, K. P., Štih, V., Aimon, S., Gjorgjieva, J., Portugues, R., & Kadow, I. C. G. (2020). Valence and state-dependent population coding in dopaminergic neurons in the fly mushroom body. *Current Biology, 30*, 2104–2115.

Silberstein, M. (2001). Converging on emergence. Consciousness, causation and explanation. *Journal of Consciousness Studies, 8*(9–10), 61–98.

Silberstein, M. (2006). In defence of ontological emergence and mental causation. In P. Clayton & P. Davies (Eds.), *The re-emergence of emergence: The emergentist hypothesis from science to religion* (pp. 203–226). Oxford University Press.

Silberstein, M. (2017). Strong emergence no, contextual emergence yes. *Philosophica, 91*, 145–192.

Simon, H. A. (1962). The architecture of complexity. *Proceedings of the American Philosophical Society, 106*, 467–482.

Simon, H. A. (1973). The organization of complex systems. In H. H. Pattee (Ed.), *Hierarchy theory. The challenge of complex systems* (pp. 71–108). George Braziller.

Skora, L. I., Yeomans, M. R., Crombag, H. S., & Scott, R. B. (2021). Evidence that instrumental conditioning requires conscious awareness in humans. *Cognition, 208*, 104546.

Slabodnick, M. M., & Marshall, W. F. (2014). Stentor coeruleus. *Current Biology, 24*(17), R783–R784.

Smith, E. S. J., & Lewin, G. R. (2009). Nociceptors: A phylogenetic view. *Journal of Comparative Physiology A, 195*(12), 1089–1106.

Smolka, J., Raderschall, C. A., & Hemmi, J. M. (2013). Flicker is part of a multi-cue response criterion in fiddler crab predator avoidance. *Journal of Experimental Biology, 216*(7), 1219–1224.

Sneddon, L. U. (2019). Evolution of nociception and pain: Evidence from fish models. *Philosophical Transactions of the Royal Society B, 374*(1785), 20190290.

Sneddon, L. U., Elwood, R. W., Adamo, S. A., & Leach, M. C. (2014). Defining and assessing animal pain. *Animal Behavior, 97*, 201–212.

Snyder, R. A. (1991). Chemoattraction of a bactivorous ciliate to bacteria surface compounds. *Hydrobiologia, 215*(3), 205–213.

Sombke, A., & Harzsch, S. (2015). Immunolocalization of histamine in the optic neuropils of *Scutigera coleoptrata* (Myriapoda: Chilopoda) reveals the basal organization of visual systems in Mandibulata. *Neuroscience Letters, 594*, 111–116.

Solvi, C., Baciadonna, L., & Chittka, L. (2016). Unexpected rewards induce dopamine-dependent positive emotion–like state changes in bumblebees. *Science, 353*(6307), 1529–1531.

Solvi, C., Gutierrez Al-Khudhairy, S., & Chittka, L. (2020). Bumble bees display cross-modal object recognition between visual and tactile senses. *Science, 367*(6480), 910–912.

Song, P. S., Kim, I. H., Rhee, J. S., Huh, J. W., Florell, S., Faure, B., & Yamazaki, I. (1991). Photoreception and photomovements in *Stentor coeruleus*. In F. Lenci, F. Ghetti, G. Colombetti, D. P. Häder, & P. S. Song (Eds.), *Biophysics of photoreceptors and photomovements in microorganisms* (pp. 267–279). Springer.

Sourjik, V., & Wingreen, N. S. (2012). Responding to chemical gradients: Bacterial chemotaxis. *Current Opinion in Cell Biology, 24*(2), 262–268.

Søvik, E., & Barron, A. B. (2013). Invertebrate models in addiction research. *Brain, behavior and evolution, 82*(3), 153–165.

Sperry, R. W. (1980). Mind-brain interaction: Mentalism, yes; dualism, no. *Neuroscience, 5*(2), 195–206.

Sperry, R. W. (1984). Consciousness, personal identity and the divided brain. *Neuropsychologia, 22*, 661–673.

Sperry, R. W. (1990). Forebrain commissurotomy and conscious awareness. In C. Trevarthen (Ed.), *Brain circuits and functions of the mind* (pp. 371–388). Cambridge University Press.

Sperry, R. W. (1991). In defense of mentalism and emergent interaction. *Journal of Mind and Behavior, 12*(2), 221–245.

Spivak, E. D. (2010). The crab *Neohelice (=Chasmagnathus) granulata*: An emergent animal model from emergent countries. *Helgoland Marine Research, 64*(3), 149–154.

Sporns, O. (2011). *Networks of the brain*. MIT Press.

Stokes, M. (1997). Larval locomotion of the lancelet *Branchiostoma floridae*. *The Journal of Experimental Biology, 200*(11), 1661–1680.

Stokes, M. D., & Holland, N. D. (1995). Ciliary hovering in larval lancelets (=Amphioxus). *Biological Bulletin, 188*(3), 231–233.

Strausfeld, N. J. (2009). Brain organization and the origin of insects: An assessment. *Proceedings of the Royal Society B: Biological Sciences, 276*(1664), 1929–1937.

Strausfeld, N. J. (2013). *Arthropod brains: Evolution, functional elegance, and historical significance*. Harvard University Press.

Strausfeld, N. J. (2020). Nomen est omen, cognitive dissonance, and homology of memory centers in crustaceans and insects. *Journal of Comparative Neurology, 528*(15), 2595–2601.

Strausfeld, N. J., Mok Strausfeld, C., Loesel, R., Rowell, D., & Stowe, S. (2006). Arthropod phylogeny: Onychophoransn brain organization suggests an archaic relationship with a chelicerate stem lineage. *Proceedings of the Royal Society B: Biological Sciences, 273*(1596), 1857–1866.

Strausfeld, N. J., Wolff, G. H., & Sayre, M. E. (2020). Mushroom body evolution demonstrates homology and divergence across Pancrustacea. *Elife, 9*, 1–46.

Storch, V., & Ruhberg, H. (1977). Fine structure of the sensilla of *Peripatopsis moseleyi* (Onychophora). *Cell and Tissue Research, 177*(4), 539–553.

Storch, V., & Ruhberg, H. (1993). Onychophora. In F. W. Harrison & M. E. Rice (Eds.), *Microscopic anatomy of invertebrates* (pp. 11–56). Wiley-Liss.

Sulston, J. E. (1983). Neuronal cell lineages in the nematode *Caenorhabditis elegans*. *Cold Spring Harbor Symposia on Quantitative Biology, 48*(Pt 2), 443–452.

Sulston, J. E., & Horvitz, H. R. (1977). Post-embryonic cell lineages of the nematode, *Caenorhabditis elegans*. *Developmental Biology, 56*(1), 110–156.

Summers, P. J., Layne, R. M., Ortega, A. C., Harris, G. P., Bamber, B. A., & Komuniecki, R. W. (2015). Multiple sensory inputs are extensively integrated to modulate nociception in *C. elegans*. *Journal of Neuroscience, 35*(28), 10331–10342.

Swinderen, B. V. (2005). The remote roots of consciousness in fruit-fly selective attention? *Bioessays, 27*(3), 321–330.

Takemura, S. Y., Aso, Y., Hige, T., Wong, A., Lu, Z., Xu, C. S., Rivlin, P. K., Hess, H., Zhao, T., Parag, T., Berg, S., Huang, G., Katz, W., Olbris, D. J., Plaza, S., Umayam,

L., Aniceto, R., Chang, L. A., Lauchie, S., Ogundeyi, O., . . . Scheffer, L. K. (2017). A connectome of a learning and memory center in the adult Drosophila brain. *eLife, 6*, e26975.

Talkington, M. (2001). Sensation on a small scale. *Pain Research Forum.* https://www .painresearchforum.org/news/3822-sensation-small-scale

Tartar, V. (1961). *The biology of Stentor.* Pergammon Press.

Teller, P. (1992). Subjectivity and knowing what it's like. In A. Beckermann, H. Flohr, and J. Kim (Eds.), *Emergence or reduction? Essays on the prospects of nonreductive physicalism* (180–200). Walter de Gruyter.

Thompson E. (2007). *Mind in life: Biology, phenomenology and the sciences of mind.* Cambridge, MA: Harvard University Press.

Thompson, E. (2022). Could all life be sentient? *Journal of Consciousness Studies, 29*(3–4), 229–265.

Tomsic, D., & Maldonado, H. (2014). Neurobiology of learning and memory of crustaceans. In C. Derby & M. Thiel (Eds.), *Crustacean nervous systems and their control of behavior (third volume of a ten-volume set on the natural history of crustaceans)* (chapter 19, pp. 509–534). Oxford University Press.

Tomsic, D., Sztarker, J., Berón de Astrada, M., Oliva, D., & Lanza, E. (2017). The predator and prey behaviors of crabs: From ecology to neural adaptations. *Journal of Experimental Biology, 220*(13), 2318–2327.

Tononi, G. (2008). Consciousness as integrated information: A provisional manifesto. *Biological Bulletin, 215*(3), 216–242.

Tononi, G. (2015). Integrated information theory. *Scholarpedia, 10*(1), 4164.

Tononi, G. (2017). Integrated information theory of consciousness: Some ontological considerations. In S. Schneider & M. Velmans (Eds.), *The Blackwell companion to consciousness* (pp. 621–633). John Wiley & Sons.

Tononi, G., & Edelman, G. M. (1998). Consciousness and complexity. *Science, 282,* 1846–1851.

Tononi, G., & Koch, C. (2015). Consciousness: Here, there and everywhere? *Philosophical Transactions of the Royal Society B: Biological Sciences, 370*(1668), 20140167.

Tononi, G., Boly, M., Massimini, M., & Koch, C. (2016). Integrated information theory: From consciousness to its physical substrate. *Nature Reviews Neuroscience, 17*(7), 450–461.

Tracey, W. D., Jr. (2017). Nociception. *Current Biology, 27*(4), R129–R133.

Trinh, M. K., Wayland, M. T., & Prabakaran, S. (2019). Behavioural analysis of single-cell aneural ciliate, *Stentor roeseli,* using machine learning approaches. *Journal of the Royal Society Interface, 16*(161), 20190410.

Tye, M. (2000). Qualia. *The Stanford encyclopedia of philosophy.* Fall 2021 ed. Edward N. Zalta, Ed. https://plato.stanford.edu/archives/fall2021/entries/qualia/

Tye, M. (2002). *Consciousness, color, and content.* MIT Press.

Van Gulick, R. (2001). Reduction, emergence and other recent options on the mind–body problem: A philosophical overview. *Journal of Consciousness Studies, 8,* 1–34.

Van Gulick, R. (2019). Emergence and consciousness. In S. Gibb, R. F. Hendry, & T. Lancaster (Eds.), *The Routledge handbook of emergence* (pp. 215–224). Routledge.

Van Gulick, R. (2021). Consciousness. In Edward N. Zalta (Ed.),*The Stanford encyclopedia of philosophy.* Winter 2021 ed. https://plato.stanford.edu/archives/win2021/entries/consciousness/

Van Houten, J. L., Yang, W. Q., & Bergeron, A. (2000). Chemosensory signal transduction in Paramecium. *Journal of Nutrition, 130*(4), 946S–949S.

Van Kranendonk, M. J., Deamer, D. W., & Djokic, T. (2017). Life springs. *Scientific American, 317,* 28–35.

Velmans, M. (1991). Consciousness from a first-person perspective. *Behavioral and Brain Sciences* 14, 702–719.

Velmans, M. (2009). *Understanding consciousness.* Routledge.

Vision, G. (2017). Emergentism. In M. Velmans & M. Schneider (Eds.), *The Blackwell companion to consciousness* (pp. 337–348). Wiley–Blackwell.

Vintiadis, E. (2013a). Emergence. *Internet Encyclopedia of Philosophy.* https://iep.utm.edu/emergenc/

Vintiadis, E. (2013b). Why a naturalist should be an emergentist about the mind. *SATS, 14*(1), 38–62.

Vizueta, J., Escuer, P., Frías-López, C., Guirao-Rico, S., Hering, L., Mayer, G., . . . & Sánchez-Gracia, A. (2020). Evolutionary history of major chemosensory gene families across Panarthropoda. *Molecular Biology and Evolution, 37*(12), 3601–3615.

Wadhams, G. H., & Armitage, J. P. (2004). Making sense of it all: Bacterial chemotaxis. *Nature Reviews Molecular Cell Biology, 5*(12), 1024–1037.

Walter, S., & Heckmann, H.-D. (2003). *Physicalism and mental causation.* Imprint Academic.

Wells, M. J. (1963). Taste by touch: Some experiments with octopus. *Journal of Experimental Biology, 40*(1), 187–193.

Wells, M. J., & Young, J. Z. (1969). The effect of splitting part of the brain or removal of the median inferior frontal lobe on touch learning in octopus. *Journal of Experimental Biology, 50*(2), 515–526.

Wertz, A., Rössler, W., Obermayer, M., & Bickmeyer, U. (2006). Functional neuro-anatomy of the rhinophore of *Aplysia punctata. Frontiers in Zoology, 3*(1), 1–11.

White, J. G., Southgate, E., Thomson, J. N., & Brenner, S. (1986). The structure of the nervous system of the nematode *Caenorhabditis elegans. Philosophical Transactions of the Royal Society B: Biological Sciences, 314*(1165), 1–340.

Wicht, H., & Lacalli, T. C. (2005). The nervous system of amphioxus: Structure, development, and evolutionary significance. *Canadian Journal of Zoology, 83*(1), 122–150.

Williams, R. (2019). Single-celled organism appears to make decisions. *The Scientist.* https://www.the-scientist.com/news-opinion/single-celled-organism-appears-to-make-decisions-66818

Williamson, R., & Chrachri, A. (2004). Cephalopod neural networks. *Neurosignals, 13*(1–2), 87–98.

Wimsatt, W. C. (1994/2007). The ontology of complex systems: Levels of organization, perspectives, and causal thickets. In *Re-engineering Philosophy for Limited Beings: Piecewise Approximations* (pp. 193–240). Harvard University Press.

Witham, E., Comunian, C., Ratanpal, H., Skora, S., Zimmer, M., & Srinivasan, S. (2016). *C. elegans* body cavity neurons are homeostatic sensors that integrate fluctuations in oxygen availability and internal nutrient reserves. *Cell Reports, 14*(7), 1641–1654.

Wittenburg, N., & Baumeister, R. (1999). Thermal avoidance in *Caenorhabditis elegans*: An approach to the study of nociception. *Proceedings of the National Academy of Sciences, 96*(18), 10477–10482.

Wolf, H. (2008). The pectine organs of the scorpion, *Vaejovis spinigerus*: Structure and (glomerular) central projections. *Arthropod Structure & Development, 37*(1), 67–80.

Wolff, T., Iyer, N. A., & Rubin, G. M. (2015). Neuroarchitecture and neuroanatomy of the Drosophila central complex: A GAL4-based dissection of protocerebral bridge neurons and circuits. *Journal of Comparative Neurology, 523*(7), 997–1037.

Wollesen, T., Wanninger, A., & Klussmann-Kolb, A. (2007). Neurogenesis of cephalic sensory organs of *Aplysia californica. Cell and Tissue Research, 330*(2), 361–379.

Woodruff, M. L. 2016. Bacteria and the cellular basis of consciousness. *Animal Sentience, 1*(11), 2.

Wustmann, G., & Heisenberg, M. (1997). Behavioral manipulation of retrieval in a spatial memory task for *Drosophila melanogaster. Learning and Memory, 4*(4), 328–336.

Wustmann, G., Rein, K., Wolf, R., & Heisenberg, M. (1996). A new paradigm for operant conditioning of *Drosophila melanogaster. Journal of Comparative Physiology A, 179*(3), 429–436.

Yamazaki, D., Hiroi, M., Abe, T., Shimizu, K., Minami-Ohtsubo, M., Maeyama, Y., & Tabata, T. (2018). Two parallel pathways assign opposing odor valences during Drosophila memory formation. *Cell Reports, 22*(9), 2346–2358.

York, C. A., & Bartol, I. K. (2016). Anti-predator behavior of squid throughout ontogeny. *Journal of Experimental Marine Biology and Ecology, 480*, 26–35.

Young, J. Z. (1961). Learning and discrimination in the octopus. *Biological Reviews, 36*(1), 32–95.

Young, J. Z. (1971). *The anatomy of the nervous system of Octopus vulgaris.* Oxford University Press.

Young, J. Z. (1972). The organization of a cephalopod ganglion. *Philosophical Transactions of the Royal Society of London. Series B, Biological Sciences, 263*(854), 409–429.

Young, J. Z. (1974). The central nervous system of Loligo I. The optic lobe. *Philosophical Transactions of the Royal Society of London. B, Biological Sciences, 267*(885), 263–302.

Young, J. Z. (1991). Computation in the learning system of cephalopods. *Biological Bulletin, 180*(2), 200–208.

Young, J. Z. (1995). Multiple matrices in the memory system of Octopus. In J. N. Abbott, R. Williamson, & L. Maddock (Eds.), *Cephalopod neurobiology* (pp. 431–443). Oxford University Press.

Zachariou, V., & Carr, F. (2014). Nociception and pain: Lessons from optogenetics. *Frontiers in Behavioral Neuroscience, 8*, 69.

Zullo, L., & Hochner, B. (2011). A new perspective on the organization of an invertebrate brain. *Communicative & Integrative Biology, 4*(1), 26–29.

Zullo, L., Sumbre, G., Agnisola, C., Flash, T., & Hochner, B. (2009). Nonsomatotopic organization of the higher motor centers in octopus. *Current Biology, 19*(19), 1632–1636.

Zumberge, J. A., Love, G. D., Cárdenas, P., Sperling, E. A., Gunasekera, S., Rohrssen, M., & Summons, R. E. (2018). Demosponge steroid biomarker 26-methylstigmastane provides evidence for Neoproterozoic animals. *Nature Ecology & Evolution, 2*(11), 1709–1714.

Index